Biotechnology Monographs

Volume 1

Editors

S. Aiba · L. T. Fan · A. Fiechter · K. Schügerl

Israel Goldberg

Single Cell Protein

With 32 Figures

Springer-Verlag
Berlin Heidelberg New York Tokyo

Israel Goldberg

Department of Applied Microbiology
Institute of Microbiology
The Hebrew University –
Hadassah Medical School
P.O. Box 1172, Jerusalem 91010
Israel

ISBN-13: 978-3-642-46542-0 e-ISBN-13: 978-3-642-46540-6
DOI: 10.1007/978-3-642-46540-6

Library of Congress Cataloging in Publication Data.
Goldberg, Israel, 1943– . Single cell protein. (Biotechnology monographs; vol. 1)
Bibliography: p. Includes index. 1. Single cell proteins. I. Title. II. Series. [DNLM: 1. Dietary Proteins. 2. Food Supply. QU 55 G618s]. TP248.S54G65 1985 664'.64 85-2813.

ISBN-13: 978-3-642-46542-0

2152/3140-543210

Preface

In early 1973, I returned to Israel from a post-doctoral fellowship at Harvard University, and was accepted as a lecturer in the Department of Applied Microbiology at the Hebrew University of Jerusalem. Shortly after my return, Professor Richard Mateles, who at that time was head of the Department, suggested that I purchase a good and comprehensive book on single cell protein (SCP) in order to expand my general knowledge in the subject I had started then to work on; that was microbial utilization of one-carbon (C_1) compounds. Naturally, I took his advice (after all, he was the Boss) and bought the book, which was the only general book published on this subject at that time, and was based on papers presented at the First International Conference on Single Cell Protein, held at the Massachussetts Institute of Technology (M.I.T.), on October 1967 (Mateles and Tannenbaum, editors) [1]. Through this book I became acquainted with the world's hunger problem that existed in the past, and ways in which it was to be solved by SCP products prepared from CO_2, fossil-based raw materials, and from wastes.

Eighteen years have passed since the M.I.T. meeting and, unfortunately, SCP did not provide the world's growing population with an abundant supply of high-quality protein. Obviously, the food hunger problem, especially in the developing countries, has not yet emerged as an issue sufficiently pressing to stimulate the world as a whole into action. Thus today, large-scale processes for SCP production are not widespread, mainly because of economic, and not because of technological reasons. Nevertheless, the commercial development of SCP during the past two decades, together with the accompanied basic and applied research, tremendously influenced modern fermentation technology, and more general biotechnology. The object of this monograph is, therefore, to describe the achievements in SCP production and the valuable contributions of SCP research and development.

The monograph is intended for all who feel the need for some first orientation in this vast and fascinating field. It aims to give the newcomer an overview of the field without confusing him with too many specifics and thus, to equip him for consulting more specialized scientific literature. Also, the monograph provides details of current basic and applied work which, hopefully will intensify the imagination and innovative skills of all those involved in development of SCP production processes.

This monograph would never have been written without the initiative of Springer-Verlag and the editorial committee of "Biotechnology Monographs". I have been fortunate in having colleagues willing and able to help and advise me during the writing of the book. It is a pleasure to acknowledge the contribution of R. I. Mateles from Stauffer Chemical Company, G. L. Solomons from Ranks, Hovis McDougall Research Limited, L. T. Fan from Kansas State University, B. A. Prior from the University of the Orange Free State, and J. S. Rokem and S. Vecht from the Hebrew University, who read and criticized in detail the bulk, or parts, of the manuscript, and who made many suggestions which have been incorporated into the monograph.

I want to express my sincerest thanks to my family, who patiently endured the many evenings and weekends that I devoted to writing and who gave encouragement when it was most needed.

Jerusalem, May 1985 — Israel Goldberg

Table of Contents

Introduction

Since the mid-1950's, there has been much emphasis on world's food shortage (world's hunger crisis) and possible ways to reduce it. A growing awareness for the acute food needs of the world's expanding population, especially in the developing countries, has led to the examination of a variety of methods to enhance agricultural and animal food sources and to produce "unconventional" food sources (microbial protein) as potential additions to the world's food supply (see the reviews of Altschul [2], Brown [3], Munro [4], Scrimshaw [5], Borgstrom [6], Litchfield [7], Solomons [8], and Mellor and Adams [9]).

The most important factor contributing to world's hunger is the accelerated population growth rate. World's population is growing faster than ever, and to describe this rate, the term "population explosion" was coined (Fig. 1). Although growth takes place everywhere, the population growth rate of the developing countries (Asia, Africa, and Latin America) is double that of the developed coun-

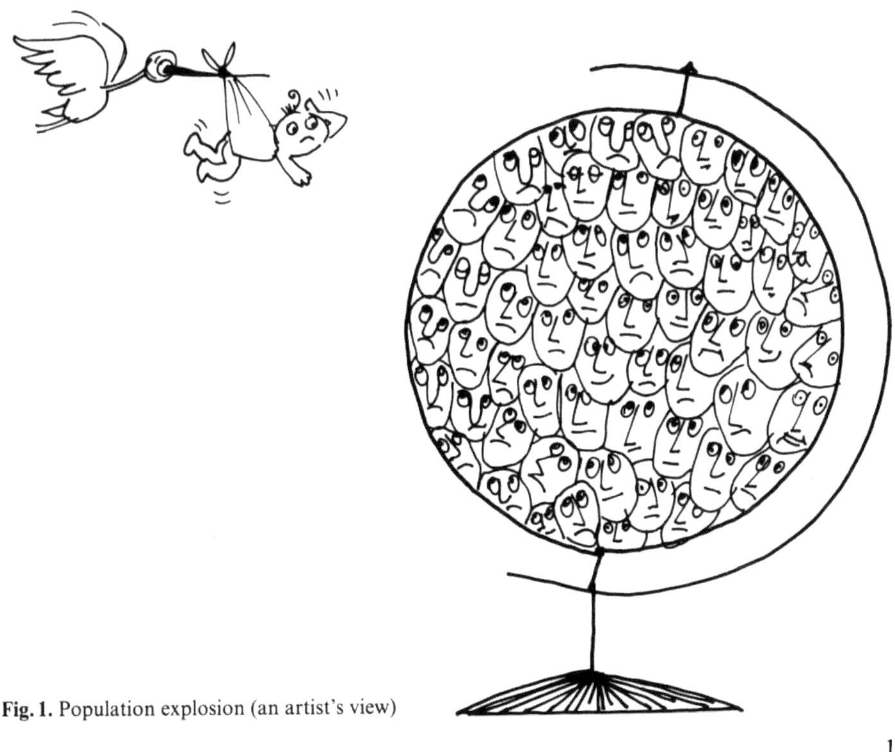

Fig. 1. Population explosion (an artist's view)

tries. According to a recent report issued by the World Bank (Mexico, August 1984) the present world's population of 4 billion people will be doubled in just 45 years, with the highest increase in population in the developing countries.

Between 1961 and 1980, food production in the developing world (including 105 countries) increased at an average annual rate of 2.6%. This was slightly greater than the corresponding average annual population growth rate of 2.4%. However, on a per capita basis, food production in the developing countries as a whole increased only 0.2% a year. Moreover, in countries like North Africa, the Middle East, and Sub-Saharan Africa, the annual growth of major food crop production was lower (between 1.7 to 2.4%) than the annual population growth (about 2.8%) [9].

In the developed countries there is ample production of major food commodities such as corn, wheat, soya, potatoes, milk, fish and eggs. These countries provide about one third of the world population with nutritious food. In developing countries, comprising large parts of the world, the opposite is true. In these countries, the inadequacies are not only in the low quantity of food, but also in its quality. The diets of people in these countries are generally low in calories and unbalanced in other nutrients, because the bulk of the diet consists mainly of cereals (mostly rice) and starchy food. Owing to the lack of variety in the diet, and particularly to the shortage of foods such as milk, eggs, fish and meat, this diet fails to provide adequate quantities of vitamins, minerals and especially high-quality protein, such that the food problem is largely a protein deficit [6].

About two-thirds of the world's population live in squalid unhygenic conditions which enhance sickness. This factor in conjunction with a poor diet, effects gross malnutrition. It is estimated that presently more than 450 million people, most of whom are living in developing countries, are hungry and undernourished [9]. Recently, an increasing demand for high-quality protein has been observed in developed countries, as overall living standards have improved. Concomitantly, in the developing countries, the rising per capita income has further contributed to the ever-increasing demand for protein. The developed countries have the funds and technical expertise to cope with an increased domestic demand for food, and thus have improved the level of protein production, thereby widening the food production gap, more accurately called "the protein hunger gap", between the "have" and "have not" countries [3, 6].

The principal methods of maximizing food availability in the developing countries are by enlarging the area of cropped land, increasing crop yields, and by maximizing imports of surplus commodities from the developed countries [9]. It is highly unrealistic to expect developing countries to raise agricultural production to that of developed countries, as these advanced countries have superior resources in soils, forests, fertilizers, pesticides, agricultural equipment, water, and above all, capital. Rising food transportation costs suggest that food export arrangements from developed countries are temporary measures which cannot be relied upon indefinitely to furnish substantial or continuing relief to needy areas [6]. Presently, even with the efforts of most of the developing nations, together with help from international agencies, developed countries and private philanthropic foundations, production (or importation) of food is failing to keep in pace with the population growth rates. It would appear, therefore, that the hunger crisis in developing countries is of such dimensions that there is no single remedy that could resolve the inherent difficulties. Consequently, the application of science and technology towards the development of "unconventional" sources of protein is vital as an adjunct to all-out efforts to lower the current deficit in food and feed (mainly protein) in many developing countries. Many of these countries produce an excess of materials high in carbohydrates. These can be utilized in fermentation processes to produce microbial protein, which, in turn, can be used to upgrade both foods and feeds. Since microorganisms need food (carbon compounds) for growth, they may be competing partially for man's resources. This can be avoided in many cases by using the ubiquitous organic waste products as carbon sources for microbial growth.

Borgstrom [6] wrote that microbes of the future era will be called "live-stock" because in principle they are converters of organic matter, as are domestic animals. Microbes have two significant economic advantages over animals:

they grow faster and
they utilize simpler food.

The greater speed and efficiency of microbial protein production compared with animal protein production may be illustrated as follows: protein (dry weight) production from 1000 g feed amounts to 49 g for chicken (doubling time, t_D, is 3 weeks), 42 g for pigs ($t_D = 8$ weeks), 14 g for bullock ($t_D = 25$ weeks), and 136 g for fungal protein ($t_D = 6$ h) intended for human food.

Microorganisms have several other specific advantages over plants and animals, with whom they compete as a potential food or feed source. Microorganisms and plants are capable of producing protein from inorganic nitrogen; plants need an entire season to grow, whereas microorganisms double their cell mass within a few hours. Furthermore, the protein content of the microorganisms is high and microorganisms can easily be modified genetically to produce cells of desired nutritional content. Their production can be based upon raw materials readily available in large quantities that would otherwise be wasted, and may have a cost of disposal. Some microorganisms can be grown in continuous processes and can then be collected by simple harvesting techniques, independent of climatic variations, and their production plants require only a small land area (though it should be noted that in developing countries the problem of land area availability does not compare with that of finding the trained personnel). Finally, the microbial protein can be prepared almost anywhere in the world and gives an homogeneous product of defined composition and quality.

These considerations combined, lend credibility to the research aimed at developing feasible means of large scale production and utilization of protein from single cells (single cell protein) which is safe, nutritious, acceptable, and within the economic resources of the developing countries.

The term "single-cell protein" (SCP) was coined at M.I.T. in May, 1966, by Professor Carroll Wilson [10]. This term refers to the dried cells of microorganisms such as bacteria, yeasts, molds, algae, actinomycetes and higher fungi grown in large-scale culture systems for use as protein sources in human foods or animal feeds. Other terms had been suggested, i.e., in France the Medical Academy proposed the term "protéines de biosynthèse" and in Japan and Italy the biomass obtained by fermentation from hydrocarbons was known as "petroprotein" and "bioprotein," respectively. These terms blur the distinction between microbial cells as such, which may contain from 35 to 70% protein, and protein concentrates, extracted from them. The protein can be consumed directly as part of the cell itself, particularly in animal feed formulations, or can be extracted and processed into fibers or meat-like items to produce nutritious human food. Some microbiologists prefer to call products consisting of whole cells "microbial biomass" or "microbial biomass products" and proteins, protein extracts, or protein concentrates obtained from these cells as "single cell protein". However, this distinction will not be retained in the present monograph. Here, single cell protein will refer to all types of single-cell proteinaceous matter.

Single-cell microorganisms have been an important component of human foods for thousands of years. Since ancient times mankind has used yeasts to change fruit juice into wine and to make bread rise [6]. People recovered top fermenting yeasts, mostly *Saccharomyces cerevisiae*, from the production of fermented beverages such as beer and wine, and then used these yeasts as a leavening agent in baked foods (Fig. 2). Other harmless microorganisms were used in man's food production; lactic acid bacteria were used in making fermented milks, cheeses and sausages, and molds were used in production of a variety of Oriental fermented foods. The use of protein and vitamin containing yeasts (or other microbial biomass) directly for food instead of baking and brewing is a twentieth-century concept [6].

This transition began in the 19th century (1860) when the Vienna process for baker's yeast production was introduced. The process was further developed, when, in 1879, continuous aeration of the vats used for producing baker's yeast was introduced in England. Also, in 1900, centrifugation was introduced in the USA as a means for separating yeast cells from the culture broth. Through these technologies, the basis of an efficient process for microbial biomass production was established.

Modern technology for SCP production originated in Germany during World War I when *Saccharomyces cerevisiae* was grown on molasses for consumption as a protein supplement. The Germans also grew *Candida utilis* (*Torula* yeast) in hardwood liquors from the pulp and paper industries; such plants were in full-scale production at the outbreak of World War II (Fig. 2). About 15,000 tons of *Candida* yeast were thereby added each year as a meat substitute and meat extender first for the army and prisoners-of-war, and later for the civilian population [12]. Food yeast was also incorporated into rations by the Russians and Japanese during World War II. In 1920 (in West Germany), *Aspergillus fumigatus* was grown on straw supplemented with inorganic nitrogen fertilizer for animal feed, and other fungi were grown by the submerged culture technique and fed to human populations. Later, the Waldhof fermentor was being developed which provided both agitation and aeration, and represented a significant technological advance for mass cultivation of microbial cells. In addition, a great deal of our knowledge about microbial cell production came from the mass cultivation of molds and fungi during the U.S. development of aerobic fermentations for the production of citric acid, gluconic acid, fumaric acid and penicillin and, hence, the advent of continuous processes for growing *S. cerevisiae* on a large scale in

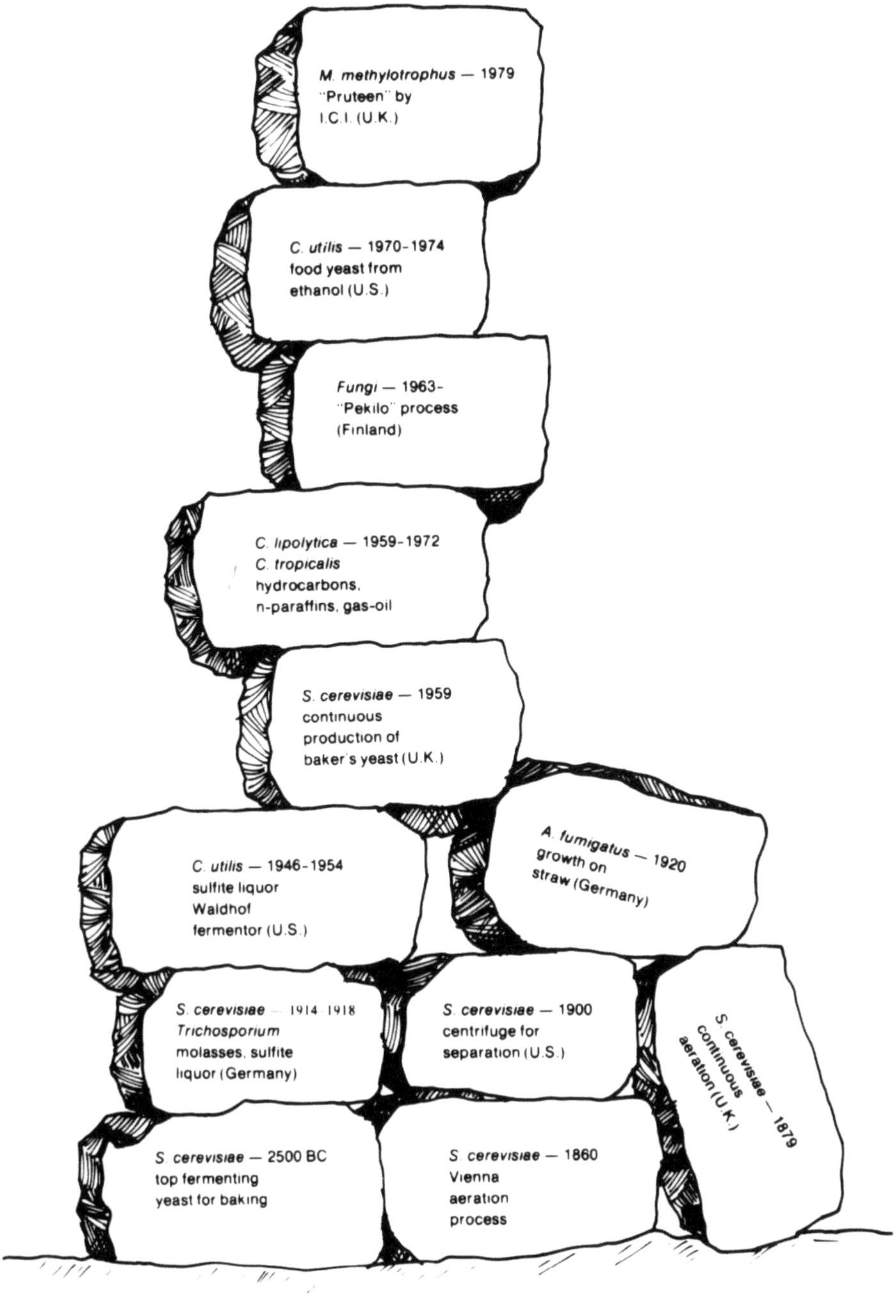

Fig. 2. Developments in microbial cell production. (Based on Litchfield [11])

England. After World War II, many processes were developed for SCP manufacture [11, 12]. Most notable was the introduction of *Candida* yeast production into the USA and into Europe which has continued until the present time. In 1977, 53,000 and 160,000 tons of dried food and fodder yeasts were produced in North America and Europe, respectively, as compared with the world's production of Baker's yeast of about 187,700 tons [13].

Since the 1960's the number of studies on SCP production have multiplied due to a combination of several factors. The first and major factor concerned the economic returns which were to be expected from the selling of a relatively cheap, yet, a high-quality protein source (to needy consumers). Other factors were based on political reasons, shortage of foreign exchange, and scientific curiosity in developing a new technology (which at that time, was thought to be very useful for the coming future). Another factor was the demonstration by A. Champagnat (see below), of the successful cultivation of microorganisms on petroleum fractions; a process which efficiently transforms various inexpensive carbon sources into good quality microbial protein. At that time, the major oil companies became interested in this observation mainly because of the economic reasons mentioned above. Their investments in SCP production provided them good public relations due to the humanitarian aspects of supplying food for the exploding population of the developing countries. Lastly, the increasing public expectation of the quality of the environment in developed countries (and to a lesser extent in developing countries) had inforced the need to provide proper and quick solutions for the disposal of organic wastes of high chemical oxygen demand.

Perhaps the most significant breakthroughs in SCP production are those associated with studies on the growth of *Candida* species for animal feed applications. Typical examples are *C. lipolytica* and *C. tropicalis* grown on a variety of petroleum-based hydrocarbon substrates including both n-paraffins and gas-oil. The introduction of hydrocarbons as a substrate for SCP manufacture brought with it extensive publicity and a considerable amount of microbiological and biochemical research [14]. Along with these studies, between 1959 and 1972, technological problems in scaling-up SCP production processes were solved. Airlift fermentors were first introduced on a commercial scale by British Petroleum (BP). This was for the production of *Candida* yeast from gas-oil at their pilot plant at Lavera, France [11]. However, in 1973, the era of low-priced and easily available petroleum (or fossil-based raw materials) came to a close, mainly due to changes in relations between petroleum producing and consuming countries [15]. In search of cheap and plentiful raw materials, the fermentation industry shifted toward natural renewable resources, referred to as biomass, or products obtained from these resources. This shift is exemplified by the largest mold-SCP "Pekilo" process (based on pulp and paper wastes) now being operated in Finland, the largest bacterium-SCP "Pruteen" process (based on methanol) now being operated by Imperial Chemical Industries (ICI) in England, and by the "Torutein" yeast-SCP process (based on ethanol) now being operated by Pure Culture Products in the USA (Fig. 2).

Although large-scale processes for manufacturing SCP products are technologically feasible, only a few processes are being operated to a limited capacity. The reasons are mainly economic in nature. It is apparent that the limits to the cost incurred in producing a microbial cell protein product, must compete with protein from crop plant (soybean) or animal product (fishmeal) prices. However, it should be pointed out that economic circumstances often change and thus a process that is not economically attractive today may become so. The shift of the fermentation industry to lignocellulosic raw material represents a tremendous potential in developing modified and more economical SCP processes. Lignocellulose is a renewable resource fairly well distributed and abundant in the form of crop residues (from forestry, agriculture and food processing), wastes, and byproduct effluents. Linked to this development we have to remember that there exists an economic incentive concerned with the disposal of these lignocellulosic wastes which is, otherwise, a serious pollution problem.

Based on these considerations, Rolz [16] has suggested that in order to improve the economic analysis, rather than use the "standard" SCP process, SCP should be produced as a low cost byproduct in a multiproduct microbial process. Similarly, Jarvis and Holms [17] concluded that the economic viability of single cell protein is limited unless the protein is produced as a byproduct of some other biotechnology, e. g., energy production.

This approach certainly represents a "creative endeavor" for future biotechnological research and development. However, this multiproduct process concept is not within the scope of the present monograph. The common main objective in all "standard" processes for producing microbial cell material for food and feed is to achieve maximum productivity (cell concentration × specific growth rate) and biomass yield coefficient (dry weight of cells per unit weight of substrate). At the same time, it is aimed at obtaining a product having a high protein content, a good amino acid profile, low nucleic acid content (for human food), good feeding performance, good color, flavor, aroma and texture, and no residues of toxic or carcinogenic substances. These principles, which are the essence of this monograph, are not applicable to the byproduct formation of microbial biomass during processes aimed at the production of ethanol, antibiotics, biogas, organic and amino acids, etc.

Anthony [18] wrote that "one of the major achievements in biotechnology occurred before the word biotechnology was coined. This achievement was the large-scale development of the use of microorganisms as a source of single cell protein." Indeed, SCP production is an example of an actual large-scale application of science. This has stimulated a great deal of scientific research, much of it of a seemingly basic nature, which most probably would not have occurred without the economic motivations. Thus, since the early 1960's, when the idea of SCP as a non-agricultural means of producing foods or feeds first became prominent, there has been a rapid increase in scientific investigation of the physiology, genetics, and biochemistry of organisms which may have a role in the production of SCP. The entire area concerned with the metabolism of different carbon sources (methane, methanol, n-paraffins, etc.) and with other aspects of SCP production has received a most significant impetus due to SCP's large scale industrial application. Consequently, numerous publications and reviews have been written in this field at frequent intervals by various investigators (Table 1). Most of the books on SCP contain papers presented in symposia or conferences confined to certain raw materials or to specific aspects of SCP production. Others deal with

Table 1. The production of single cell protein (SCP)

	Ref.
(a) Books on SCP	
Mateles and Tannenbaum (eds.) (1968). Single-cell protein	[1]
De Pontanel (ed.) (1972). Proteins from hydrocarbons	[19]
Gutcho (1973). Proteins from hydrocarbons	[20]
Davis (ed.) (1974). Single-cell protein	[21]
Tannenbaum and Wang (eds.) (1975). Single-cell protein	[22]
Wagner (ed.) (1975). Proteins from microorganisms	[23]
Rockwell (1976). Single cell proteins from cellulose and hydrocarbons (process details)	[24]
Gaden and Humphrey (eds.) (1977). Single cell protein from renewable and nonrenewable resources	[25]
Rose (ed.) (1979). Microbial biomass	[14]
Shelef and Soeder (eds.) (1980). Algae biomass production and use	[26]
Harrison, Higgins and Watkinson (eds.) (1980). Hydrocarbons in biotechnology	[27]
Ferranti and Fiechter (eds.) (1983). Production and feeding of single cell protein-lignocellulose and whey	[28]
(b) Selected general reviews on SCP	
Lipinsky and Litchfield (1970). Algae, bacteria and yeasts as food or feed	[29]
Snyder (1970). Microbial sources of protein	[30]
Kihlberg (1972). The microbe as a source of food	[31]
Waslien (1975). Unusual sources of proteins for man	[32]
Litchfield (1977). Comparative technical and economic aspects of SCP processes	[33]
Rivière (1977). Microbial proteins	[34]
Cooney et al. (1979). Single cell protein: engineering, economics and utilization in foods	[35]
Mateles (1979). The physiology of SCP production	[36]
Hamer and Hamdan (1979). Protein production by microorganisms	[37]
Hamer amd Harrison (1980). SCP: the technology, economics and future potential	[38]
Reed (1982). Microbial biomass and single-cell protein	[39]
Litchfield (1983). Single-cell proteins	[7]
Samuelov (1983). Single-cell protein production: review of alternatives	[40]
Solomons (1983). Single cell protein, an industrial and commercial viewpoint	[8]
Batt and Sinskey (1984). Use of biotechnology in the production of SCP	[41]

specific raw materials such as CO_2, lignocellulose, whey and hydrocarbons. The same is true for the many review articles published which emphasize different aspects of this vast subject. The need for a book describing different carbon substrates and approaches for SCP production, as well as basic biochemical, microbiological and physiological problems observed in biomass production, will hopefully be fulfilled by the present monograph.

In the early years, the companies that were involved in SCP production devoted very large programs primarily to its engineering aspect, and some reached pilot plant programs rather quickly. However, it soon became apparent that the biological problems had been neglected, and in many cases the SCP programs were halted, reduced in scope or re-examined. One consequence of this re-examination is that much useful work has been done by industrial groups and scientists in universities or other research institutes on many biological aspects of SCP production. The second consequence is a much broader one and relates to the contributions that SCP research and development programs (engineering and biolog ical studies) have made to the overall development of fermentation technology or, more generally, to the development of modern biotechnology. This monograph is aimed at describing these contributions as well as current achievements in SCP production, and providing a useful guide for those who are interested in this fascinating subject but who are overwhelmed by the vast literature in this field. Also, it is hoped that this monograph will provide a stimulus for further innovative research and development in the production of single-cell protein.

In this short monograph it is not possible to review all aspects of such a comprehensive subject, nor to describe details of all studies made in each of various specific aspects. Therefore, only selected works are mentioned in order to concentrate on principal approaches and innovative ideas in research and development of SCP products.

1 Organisms and Substrates

1.1 Introduction

1.1.1 Microorganisms Used in SCP Production

Microorganisms suitable for single cell protein production are divided into four main categories: bacteria, yeasts, molds/fungi and algae. During the last forty years, research has concentrated on several strains of microorganisms which are of commercial importance. It will be useful to list the chief characteristics that are desirable in microorganisms to be used for production of SCP, as shown in Table 2.

Bacteria

The most important of these characteristics, from the viewpoint of an efficient culture system for the production of SCP, is the efficiency of conversion of substrate carbon to cell carbon, or the growth yield (Table 2). This is important not only because a major part of the operating cost in SCP production is that of the carbon substrate, but also because the growth yield has important implications

Table 2. Characteristics desirable in microorganisms to be used in SCP production

• High growth rate • Growth on simple media – no requirement for expensive growth factors • High yield coefficient on a carbon substrate • High affinity for the carbon substrate • Ability to utilize complex substrate and/or mixture of carbon substrates when required • Ability to grow at high cell densities • Resistance to substrate and/or product toxicity • Stable growth in continuous culture • High optimum growth temperature • Stable fermentation – resistance to contamination • Capability of genetic modification • Use ammonia as nitrogen source • pH tolerant	← Physiological characteristics Other requirements ↓ • Protein, fat and carbohydrate content of high quality • Low nucleic acid content • High nutrient content • High digestibility • Absence of toxicity • Good taste • Ease of recovery • Amenable to further processing, e.g., drying, without changes of color, texture, smell, etc.

for the production process. A higher conversion efficiency results in a higher overall cell yield and productivity, lower oxygen requirement per unit of cell produced, and lower amount of heat evolved per unit of cell produced. This, in turn, will affect the cost of aeration and cooling of the fermentors which constitute a large proportion of the operating costs (see below). In this respect the use of a bacterium, rather than a yeast, as the process organism offers some attractive advantages. Firstly, growth rate and cellular yield from carbon sources are higher for bacteria than other organisms. Secondly, bacteria in general have higher protein contents than other organisms. Thirdly, the quality of the protein, in terms of its amino acid profile, is better in bacteria as they contain higher concentrations of sulfur amino acids and lysine. Bacteria do have some disadvantages indicated by the limited number of bacterial species that have been grown specifically for food purposes.

Other disadvantages include the very small size of bacteria, which causes difficulties during the separation stage, the relatively high content of nucleic acids, which renders their use unsuitable as a human food, the unpleasant odor caused by some bacteria, and lack of assurance that the type of bacteria chosen for mass cultivation would not include pathogenic species.

Yeasts

By far the greatest experience in the manufacture and use of microbial foods has been with yeasts. The possibility of growing yeasts as a food per se for direct human consumption was first explored by Max Delbruck and his colleagues at the Institut für Gärungsgewerbe in Berlin [42]. Thousands of tons of yeast (*Candida utilis*) were consumed in Germany during World Wars I and II as a meat substitute and meat extender for the army, prisoners-of-war, and the civilian population. At present, three types of yeast products are commonly available: brewer's yeast, dried yeast and dried *Torula* yeast. Both brewer's yeast, which is a byproduct of the beer industry, and dried yeast, are of the same genus; *Saccharomyces*. The dried *S. cerevisiae* yeast is superior to the brewer's yeast, as it does not require debittering, and has a higher content and more uniform level of vitamins. The term "food yeast" was apparently coined by the British Royal Society in 1941 to focus attention on yeast grown expressly for human consumption, as opposed to "fodder yeast"; the dried yeast obtained as a brewing industry byproduct and used as a feeding supplement for animals [42]. The *Torula* yeast or *Candida utilis* is a different strain, and is capable of assimilating almost all carbohydrates supplied for energy, including pentoses, whereas *S. cerevisiae* is not capable of utilizing pentose carbohydrates.

Yeasts are known for their low concentration of sulfur containing amino acids. This is true not only of *S. cerevisiae and C. utilis*, but has been reported for other yeast strains as well [43]. Though yeast protein is low in methionine and cysteine, it contains sufficient quantities of other essential amino acids such as lysine, tryptophan and threonine (according to the Ministry of Agriculture, Food and Fisheries, the threonine requirements are not met in the U.K. diet.).

Yeasts are a good source of the B group vitamins and also provide small amounts of vitamin E and provitamin D, but few yeast strains contain vitamin A. The quantities of these vitamins vary according to strain, specific conditions of cultivation, carbon source and methods of production. As an example, *Candida* yeast contains a high concentration of pantothenic acid which is destroyed during the final stages of processing.

Other advantages of yeasts, as compared with bacteria, are the relatively low nucleic acid content, the greater size which makes separation easier, the lower toxicity potential, and the greater acceptance by consumers.

Fungi

The use of fungi as food is not a new concept. Higher fungi, known as mushrooms, have been used as food flavoring for centuries. However, the idea of mass cultivation of fungal mycelium as a protein source is relatively new. In the 1940's, fungi were studied, mainly as potential producers of organic acids and antibiotics. The possibility of using fungi as a source of protein was neglected, with only a few exceptions [8, 44, 45]. Recent research on fungal-based single cell protein has shown that:

1) the growth rates of fungi are usually slower than those of yeasts and bacteria,
2) their protein content is usually lower than that of bacteria and yeasts,
3) fungal protein is often deficient in sulfur amino acids,
4) the problem of digestibility of their cell wall is often a serious consideration, and
5) fungi are more of an unknown entity with regard to nutrition and toxicology,

On the other hand, fungi have two important advantages:

1) most fungi produce a range of carbohydrate hydrolyzing enzymes and thus are able to grow on complex raw materials such as cellulose, hemicellulose, and starch, and
2) fungi can usually be recovered by simple filtration, which offers a significant reduction in capital and processing costs.

Algae

Algae are often thought to be an attractive source of SCP owing to their photosynthetic abilities. Due to this capacity, and the ready availability of CO_2, there is no need for the addition of carbon as an energy source. Furthermore, there are some species that are able to fix atmospheric nitrogen such as the *Cyanophyta* (blue-green algae).

Algal protein production is carried out in large open pools, natural lakes or ponds, and contamination problems caused by bacteria and protozoa are experienced in the process. Algal growth rates are lower than those of bacteria and yeasts although the separation (or collection) procedure is easier. The possibilities for producing algae in conjunction with the treatment of sewage and of growing algae in brackish water in arid zones, have been evaluated. However, these processes are not yet economic, due mainly to their low productivities. Feeding experiments on pigs and hens have shown that the biological value of cell-free algal protein is lower than that of casein, due to deficiencies in sulfur containing amino acids. However, the lysine content of algae is high, which may enable them to be utilized as a supplement to cereals, or fed in small quantities in conjunction with other protein sources. Of primary importance in the utilization of micro-algal SCP, is the stability of the cell wall. In general, the walls of both green and blue-green algae are not readily digested in the mammalian digestive system, with ruminants generally being better utilizers [46].

1.1.2 Carbon Sources for SCP Production

The traditional carbon sources for the production of SCP have been carbohydrate-containing byproducts of agriculture or industry, either in the form of readily assimilable sugars, or in the form of starch or cellulosic material. These sources are molasses, starchy grains, wood sugar from spent sulfite liquor and wood hydrolyzates and, to a lesser extent, whey.

A breakthrough in the fermentation process for large-scale production of SCP was the recognition of the universal availability of large quantities of petroleum-based raw materials. Originally, it was intended to use gas-oil (a cheap mixture of hydrocarbons) for SCP production, with simultaneous improvement of the pour point of the gas-oil. However, the emphasis shifted to other petroleum-based raw materials such as n-paraffins and synthetic methanol or ethanol, as will be discussed later.

As a result of the oil crisis at the end of 1973, petrochemical raw materials became more expensive whereas, soya and fish-meal have returned to their old and low price level, after a temporary heavy shortage and resultant increase in price. Consequently, the SCP world market prices for certain applications were higher than the prices of comparable natural products. In a search of alternative cheap and plentiful raw materials, the fermentation industry shifted from petroleum-based raw materials, to the natural renewable resources referred to as biomass. Parallel to these changes, research and development programs in effluent and waste disposal, from agriculture, agro-industry, and food-processing industries, have been directed towards the recovery of nutrients essentially through fermentation technology. Generally, the substrates have been carbohydrates, with SCP as the final product.

Many raw materials have been considered as carbon and energy sources for SCP production (Tables 3–5). These substrates can be divided into three main groups: CO_2, fossil mass (petroleum, natural gas, coal and oil shale) and renewable biomass (lignocellulose, starch and sugar). In the early days, the microbes used were *S. cerevisiae*, *S. uvarum*, *C. utilis* and, to a minor extent, *Kluyveromyces fragilis*. However, since the introduction of many other carbon sources in the early 1960's, it was necessary to use microorganisms other than those mentioned above. They are yeasts of the genera *Candida*, *Endomycopsis*, and *Kluyveromyces*, molds of the genera *Paecilomyces and Trichoderma*, and *Pseudomonas*-like bacteria as well as several algae. Dimmling [88] pointed out the preferred combinations of raw materials and microorganism which were being developed up to the year 1975. At that time, yeasts and bacteria were the most important microorganisms for the various carbon sources. This trend has changed during recent years, as is shown in Table 6 (which summarizes selected research studies during the last

Table 3. SCP processes from carbon dioxide

Organism	Production capacity/process scale	Organization	Location	Status	Ref.
Microalgae and prototrophic bacteria	110-liter, pilot plant	University of Tübingen	W. Germany	?	[47]
Chlorella sp.	700 t/y, pond area 83,400 m^2	Taiwan Chlorella Manufacturer Co., Ltd.	Taipei	Operating	[48]
Spirulina geitleri (or maxma)	320 t/y, pond area 900 m^2	Sosa Texcoco, S.A.	Mexico	Operating	[49, 50]
Scenedesmus acutus	200 t/y	Kohlenstoffbiologische Forschungsstation e. V. Dortmund	W. Germany	?	[51]
Spirulina platensis	Pond area 400 m^2	Ben Gurion University of the Negev, Sede Boker	Israel	Operating	[52]
Micractinium pusillum, Scenedesmus dimorphus	Pond area 2270 m^2	Technion, Haifa	Israel	Operating	[53]
Dunaliella bardawil	140 t/y, pond area 20,000 m^2	Koor Chemicals, Haifa	Israel	Operating	[54]

Table 4. Raw materials from fossil mass used as carbon and energy sources in SCP processes

Raw material	Organism	Production capacity/process scale	Organization	Location	Status	Ref.
Methanol	*Methylophilus methylotrophus*	50,000–75,000 t/y "Pruteen"	Imperial Chemical Industries (ICI)	Billingham, UK	Operating	[7, 55]
Methanol	*Methylomonas clara*	1000 t/y "Probion"	Hoechst-Uhde-Gelsenberg	Frankfurt, W. Germany	Discontinued	[56]
Methanol	*Methylomonas methanolica*	4500-liter "Norprotein"	Norsk Hydro and AB Marabou "Norprotein"	Sundbyberg, Sweden	Discontinued	[57, 58]
Methanol	*Pseudomonas* C	Pilot plant	Hebrew University	Jerusalem	Discontinued	[40, 51]
Methanol	Yeast, *Pseudomonas* sp.	4000–5000 t/y 20,000-liter	Mitsubishi Gas Chemical Co.	Niigata, Japan	Discontinued	[8]
Ethanol	*Candida (Torula) utilis*	7000 t/y "Torutein"	Pure Culture Products, Hutchinson	Minnesota, USA	Operating	[59]
Ethanol	Yeast	4000 t/y	Sovnaft	Kojetin, N. Moravia, Czechoslovakia	Operating	[51]
Ethanol	*Hansenula anomela*	1500 t/y	Schick-Chemie-Technik GmbH	W. Germany	Discontinued	[8]
Ethanol	*Hansenula anomela*	30 t/y	Cepsa	Huelva, Spain	Discontinued	[8, 60]

Table 4 (continued)

Raw material	Organism	Production capacity/process scale	Organization	Location	Status	Ref.
Ethanol	*Candida ethanothermophilum, Candida acidothermophilum*	100 t/y	Mitsubishi Petrochemical Co.	Japan	Operating	[8]
Kerosene and gas-oil	*Pseudomonas* No. 5401	1200 t/y 50,000 liter	Chinese Petroleum Corporation (CPC)	Taiwan	Discontinued	[61]
Crude oil	*Candida tropicalis* IIP4	Pilot plant	Institute of Petroleum	Gujarat, Assam, India	Operating	[62]
Gas-oil	Yeast	60,000 t/y	East Germany/Polish project	Schwedt, E. Germany	?	[51]
Gas-oil	*Candida tropicalis*	16,000 t/y "Toprina L"	British Petroleum (BP)	Lavera, Bouches-du-Rhone, France	Discontinued	[63]
Peat hydrolyzates	Yeast	4000 t/y		Boksitogorsk, USSR	Operating	[64]
n-Paraffins	*Endomycopsis (Candida) lipolytica* FH-H-5027	20 t/y 4000 liter	Hoechst-AG	W. Germany	Operating	[65, 66]
n-Paraffins	*Candida guilliermondii*	20,000–40,000 t/y "BVK Fermosin"	All-Union Research Institute of Protein Biosynthesis	Moscow, USSR	Operating	[7, 51]
n-Paraffins	*Candida lipolytica*	4000 t/y "Toprina G"	British Petroleum	Grangemouth, Scotland	Discontinued	[67]
n-Paraffins	*Candida lipolytica*	100,000 t/y "Toprina G"	Italprotein, BP-Anic	Sarroch, Sardinia	Discontinued	[68]
n-Paraffins	*Candida maltosa*	100,000 t/y "Liquipron" ("Kanepron")	Liquichimica Biosintesi SpA., Kanegafuchi (KCI)	Saline de Montebello, Calabria, Italy	Discontinued	[68]
n-Paraffins	*Candida tropicalis*	50,000 liter "Petroprotein"	Gulf R & D Co.	Wasco, California USA	Discontinued	[69]
n-Parafins	*Candida pichia*	60,000 t/y "Roniprot"	Dainippon & Petrom	Curtea de Argues, Jas-Syon, Romania	?	[8, 70]
n-Paraffins	*Acinetobacter cerificans*	200 liter "Protecel"	Esso-Nestle	Linden, New Jersey, USA, La Tour de la Pailz, Switzerland	Discontinued	[8]
n-Paraffins	Yeast	25 t/y	Groupment Francais des Protéines	Soleige, France	?	[51]
n-Paraffins	*Candida paraffinica*	60,000–120,000 t/y "Viton"	Dainippon Ink & Chemicals, Inc.	Tokyo, Japan	Discontinued	[51]

Table 5. Raw materials from renewable biomass used as carbon and energy sources in SCP processes

Raw material	Organism	Production capacity/process scale	Organization	Location	Status	Ref.
Food-grade glucose	*Fusarium graminearum*	50–100 t/y "Mycoprotein"	Ranks Hovis McDougall, Ltd.	High Wycombe, UK	Operating	[8, 45]
Carbohydrate waste (confectionary manufacturing effluent)	*Candida utilis*	500 t/y	Tate and Lyle R & D Ltd. and Geo Basset and Co. Ltd.	Sheffield UK	?	[71, 72]
Molasses	Yeast	20,000 t/y	Société Nationale Sempac	Algiers, Algeria	?	[51]
Potato wastes	*Saccharomyces fibuliger, Candida utilis*	10,000 t/y "Symba"	Swedish Sugar Company	Eslov, Sweden	Discontinued	[73, 74]
Agave juice	*Candida utilis*	1400 t/y	Patronato del Maguey	Mexico	Operating	[75]
Coffee wastes	*Trichoderma harzianum Rifai*	40,000 liter	ICAITI & Curazao	Guatemala and El Salvator	Operating	[11, 76]
Sulfite waste liquors	*Paecilomyces variotii*	10,000 t/y "Pekilo protein"	Tampela & Finnish Pulp and Paper Res. Inst.	Jamsankoski, Finland	Operating	[77]
Sulfite waste liquors	*Candida utilis*	5000 t/y	Rhinelander Paper Co.	Wisconsin USA	Operating	[7]
Sulfite waste liquors	*Candida utilis, Candida tropicalis*	300 liter	SFIT-ETH	Zurich, Switzerland	Operating	[78]
Sulfite waste liquors	*Candida utilis*	5000 t/y	Boise Cascade Co.	Salem, Oregon, USA	Discontinued	[79]
Sulfite waste liquors	*Candida utilis*	10,000 t/y "Silva"	Metsaluton Sellulosa Oy	Aanekoski, Finland	Operating	[8]
Whey	*Kluyveromyces fragilis*	5000 t/y "Amber Nutrex"	Amber Laboratories	Wisconsin, USA	Operating	[80]
Whey	*Penicillium cyclopium*	300 t/y	Heurty, S.A.	France	Operating	[81]
Whey	*Candida intermedia*	60,000 liter	Vienna-Unterkartner, Molkerei	Klagenfurt, Austria	Operating	[82]
Whey	*Kluyveromyces fragilis, Turolopsis bovina* and *Turolopsis sphaerica*	8000 t/y "Protibel"	Bel Fromageries Industries	Vendome, France	Operating	[83]
Whey	*Kluyveromyces fragilis*	750 t/y "Wheast"	Knudsen Milk Products Co.	Los Angeles, USA	Operating	[84]
Milk ultrafiltration permeate	*Trichosporon cutaneum*	7000 liter	Swiss Federal Institute of Milk	Zurich, Switzerland	Operating	[85, 86]
Cellulosic wastes	*Chaetomium cellulolyticum*	1000 liter	University of Waterloo	Ontario, Canada	Operating	[87]
Cellulosic wastes	*Trichoderma viride, Candida utilis*	Pilot plant	U.S. Army	Natick, USA	Operating	[51]

Table 6. Raw materials suggested for SCP production – selected research studies (1979–1984)

Raw material	Organism(s) used	Ref.
Lignocellulose:		
Lignin	*Trichosporon penicillatum* 3G	[89]
Winged bean haulm (Philippines)	*Trichoderma viride, Trichoderma koningii, Sporotrichum pulverulentum, Chaetomium cellulolyticum, fungal strains:* AM-6, SOP, B5-40-G	[90]
Cellulosic wastes	*Candida utilis* and *Saccharomyces cerevisiae*	[91]
	Chaetomium cellulolyticum	[87]
	Cellulomonas sp.	[92]
Piggery slurry-cellulosic components	*Sporotrichum pulverulentum*	[93]
Rice straw waste	*Candida utilis*	[94]
	Penicillium janthinellum	[95]
	Aspergillus terreus, Candida utilis, Candida marina, Candida tropicalis, Saccharomyces cerevisiae, Rhodotorula marina	[96]
	Trichoderma harzianum	[97]
Paper waste (Japan)	*Cellulomonas flavigena* KIST 321	[98]
Fiber board waste (Sweden)	*Sporotrichum pulverulentum* ATCC 32629	[99]
Soybean waste	*Rhodopseudomonas sphaeroides S*	[100]
Grass juice	*Saccharomyces cerevisiae*	[101]
Bagasse substrate	*Aspergillus terreus* GN1	[102]
Sugarcane bagasse	*Cellulomonas* strain II bc	[103]
	Trichoderma longibrachiatum Rifai	[104]
	Candida utilis and *Paecilomyces variotii*	[105]
Grape mast	*Candida utilis* ISS 28	[106]
Poultry waste	Mixed culture-yeast and bacteria	[107, 108]
Cow manure waste	*Spirulina maxima*	[109]
Jerusalem artichoke tubers	*Kluyveromyces* strains of *marxianus* and *fragilis, Schizosaccharomyces pombe, Debaromyces* sp.	[110]
	Several yeast strains	[111]
Leached beet cosette	*Trichoderma reesei* and *Candida marina*	[112]
Apple distillery waste	*Aspergillus niger* ATCC 6275	[113]
Whiskey distillery spent wash	*Geotrichum candidum, Hansenula anomela, Candida krusei*	[114]
Mangostone (India)	*Aspergillus niger*	[115]
Coffee pulp juice	*Aspergillus oryzae, Trichoderma harzianum, Penicillium crustosum, Gliocladium deliquescens*	[116, 117]
Banane waste	*Pichia spartinae*	[118]
Starch:		
Corn and cassava	*Candida tropicalis*	[119, 120]
	Rhodopseudomonas gelatinosa	[121]
	Schwanniomyces alluvius IGC 2829	[122]
Starch	*Lipomyces kononenkoae*	[123]
	Schwanniomyces alluvius	[124]
Tapioca starch	*Aspergillus niger*	[125]
Potato tuber fractions	*Penicillium piscarium, Penicillium simplicissimum, Aspergillus niger, Chaetomium cellulolyticum, Diplodia citricola*	[126]
	Schwanniomyces castellii	[127]
Skin of green plantain (mostly starch)	*Pichia spartinae*	[128]

Table 6 (continued)

Raw material	Organism(s) used	Ref.
Petrochemicals:		
Methanol	*Hansenula polymorpha* I-3	[129]
	Corynebacterium yamanasiensis	[130]
	Pseudomenas stuzeri, Pseudomonas aeruginosa, Micrococcus methylotrophus	[131]
	Pichia pastoris IFP 206	[132, 133]
Natural gasoline (C_5) (Kuwait)	Yeast sp.	[134]
n-Paraffins	*Candida guilliermondii*	[135]
	Candida oleophila	[136]
	Thermoleophilum album	[137]
Slack wax, hexadecane, naphtha	*Candida scotti*	[138]
Peat hydrolysates	*Bacillus polymyxa, Clostridium acetobutylicum, Candida utilis, Candida tropicalis, Candida humicola, Lipomyces lipoferus, Sporobolomyces pararoseus*	[64]
Ethanol	285 yeast strains. The best are: *Candida kruzei, Candida sorbosa, Candida guilliermondii, Trichosporon cutaneum*	[139]
Ethanol – isopropanol mixtures	*Candida kruzei*	[140]
Sugar:		
Date carbohydrates (glucose and fructose)	*Candida utilis, Saccharomyces cerevisiae, Saccharomyces rouxii, Ogi* yeast	[141]
Citric acid fermentation wastes	*Candida scottii, Candida aquatica, Candida bimundalis, Candida vartiovaarai*	[142]
Pinapple canning effluent	*Candida utilis*	[143, 144]
	Hansenula sydowiorum, Candida utilis	[145]
Hydrolysate of shell fish waste chitin	*Pichia kudriavzevii*	[146]
Lemonade processing wastewater	*Candida utilis, Saccharomyces fragilis*	[147]
Whey	Yeast strains (13 strains), *Wingea robertsii*	[148]
	Aspergillus flavus, Aspergillus niger, Aspergillus ochraceus, Aspergillus terreus	[149]
	Kluyveromyces fragilis IMAT 1872	[150, 151]
	Kluyveromyces fragilis NCYC 587, *Candida pseudotropicalis* NCYC 744	[152]
Onion waste products	*Saccharomyces* strains	[153]
Spent sulfite liquor	*Candida utilis* NRRL Y-900	[154]
	Candida tropicalis CBS94	[155]
Miscellaneous:		
Sauerkraut wastewater (lactic acid)	*Candida utilis* NRRL Y-900	[156]
Organic acids (from treated animal wastes)	*Candida ingens*	[157]
Waste fishery refrigeration brines (organic matter)	*Candida utilis*	[158]
Olive black water (lactic acid)	*Saccharomyces lipolytica*	[159]
Palm oil (triglycerides and saturated fatty acids) (Malaysia)	*Candida rugosa, Candida utilis, Candida tropicalis, Candida lipolytica*	[160]
(Japan)	*Torulopsis candida* TAMY128	[161, 162]
(Japan)	*Candida blankii* CBS 1898	[163]

Table 6 (continued)

Raw material	Organism(s) used	Ref.
Palm oil and rape-seed oil (France)	*Candida curvata, Candida rugosa, Candida deformans, Candida lipolytica, Candida parapsilosis, Cryptococcus uniguttilatus, Geotrichum candidum, Trichosporum cutaneum, Rhodotorula pilimance*	[164]
Plastic waste (polyethylene)	*Candida tropicalis*	[165]
Effluents of biogas plant	Strains of *Rhodopseudomonas*, e.g., *capsulata* . .	[166]
CO_2	*Chlorella* strain 211/8k, mixed culture MA003 . .	[167]
	Phaeodactylum tricornutum	[168]
CO_2, H_2, O_2, N_2	*Alcaligenes hydrogenophilus*	[169]
	Alcaligenes eutrophus	[170]

six years for raw material and microorganisms suggested for SCP production). Research concentrated mainly on various yeasts and fungi with only a few exceptions using bacteria. This probably reflects the inherent advantages of yeasts, and to a lesser extent, fungi, over bacteria for SCP production.

As the raw material cost accounts for a major portion (up to about 60%) of the SCP production cost (see later in the monograph), the raw material basis becomes very important. The substrate offering the greatest commercial advantages depends upon the factory location, substrate availability and competitive cost. Competitive cost means both cheaper raw materials and a better process in terms of: favorable scale of operation, higher product yield, lower energy costs and lower investment. The majority of the studies listed in Table 6 attempt either to up-grade different wastes, or to provide a protein source from supposedly cheap raw materials. Although many of the ideas might not materialize into viable commercial processes, they clearly show the diversity of thought and investigation, which has really characterized the SCP field. In particular, they do indicate that, given the right economic background, SCP might be produced on a commercial scale using different raw materials [8].

1.2 Carbon Sources and Microorganisms Used in SCP Production

1.2.1 Autotrophic Growth on CO_2

The simplest carbon source from which cell biomass can be produced is carbon dioxide. Since CO_2 is the most oxidized carbon source, its carbon has to be reduced before assimilation, and therefore, CO_2 cannot serve as an energy source. Energy is thus supplied separately in the form of light (solar) and converted to photosynthetic energy. Commercial advantage can certainly be taken where combined existence of solar energy and natural gas deposits exists.

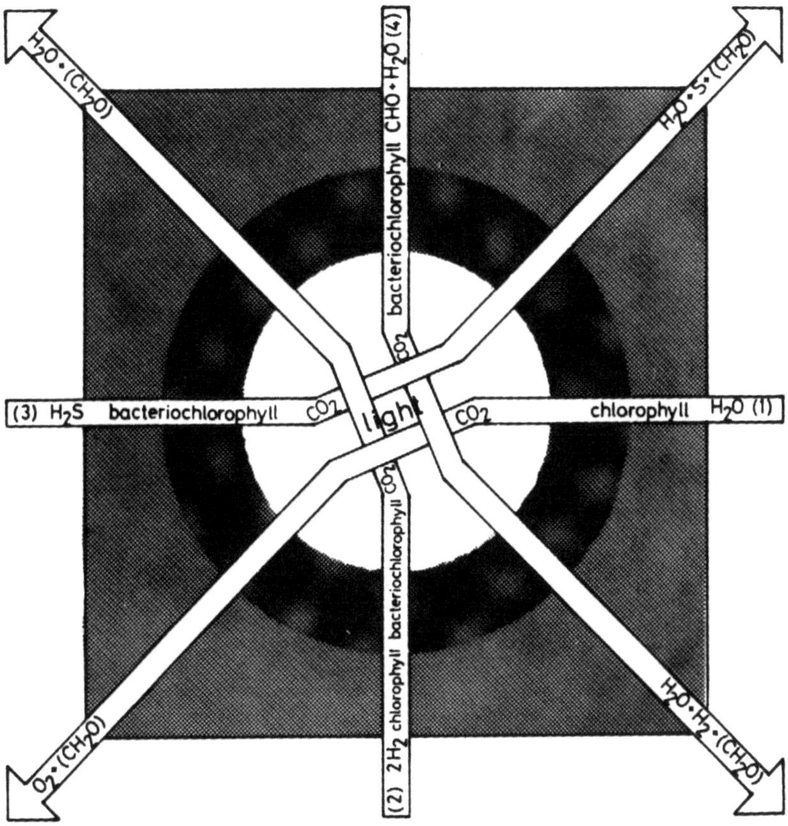

Fig. 3. Basic photosynthetic reactions of plants and microorganisms. (1) Higher plants and unicellular algae, (2) certain algae and photosynthetic bacteria, (3) photosynthetic sulfur bacteria, (4) purple phytosynthetic bacteria. (After Stanier et al. [171], Shipman et al. [172], and Reed [39])

Photosynthetic bacteria and algae have been used for SCP production from CO_2 and light energy (autotrophic growth). The main difference between the two groups of organisms is illustrated in Fig. 3. Algae, like higher plants, derive the energy required for CO_2 assimilation from sunlight or artificial light, in the range of 700 nm, by means of the chlorophyll molecules present in the cells. The photoelectron transport reactions (light reactions) of the photosynthetic system affect the light-dependent water hydrolysis required for the synthesis of reducing power. Bacterial photosynthetic processes are basically anaerobic and use molecular hydrogen, reduced sulfur compounds (such as H_2S), or some organic compounds as electron donors. Water is not used as an electron donor and molecular oxygen is never produced, contrary to that which occurs in algae. In both groups of microorganisms the CO_2 assimilation reactions (dark reaction) proceed via the Calvin pathway, described in Fig. 4.

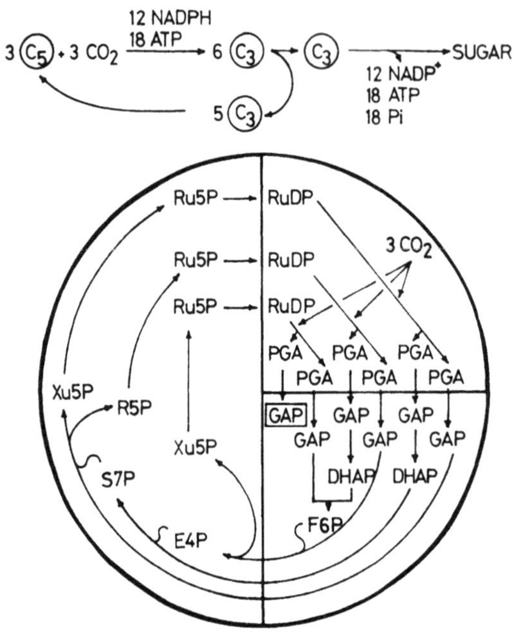

Fig. 4. Schematic representation of the Calvin pathway for CO_2 fixation. The abbreviations used are: RuDP, ribulose – 1,5-diphosphate; PGA, 3-phosphoglyceric acid; GAP, glycerate – 1,3 diphosphate; Xu5P, xylulose-5-P; R5P, ribose-5-P; S7P, sedoheptulose-7-P; E4P, erythrose-4-P; DHAP, dihydroxy-acetone phosphate. (Redrawn from Mahler and Cordes [173]). (Reprinted with permission of Harper and Row, Publishers, Inc.)

1.2.1.1 SCP from Bacterial Photosynthetic Reactions

Today, photosynthetic bacterial production of SCP is considered in only a limited number of cases [40]. Bacteria such as *Rhodopseudomonas capsulata* have been grown in Japan, with industrial waste or sewage as substrates. The bacteria grew in mixed cultures with aerobic, heterotrophic and nitrogen-fixing bacteria, yielding 1 to 2 g/liter biomass [7, 174].

Using one species of non-sulfur purple bacterium, Shipman et al. [172] grew *Rhodopseudomonas gelatinosa* on previously hydrolyzed wheat bran (30% solids) with suitable sunlight or incandescent illumination (4-watt lamps) at a pH of 7 to 7.2 and a temperature of 37 °C. A semicontinuous process was used with a retention time of about 24 h, an average productivity of 10 $kg/m^3 \cdot day$ and total production capacity of 5 tons of SCP/day. The addition of low molecular weight organic acids, carbohydrates, sewage and animal manure to the photosynthetic purple bacterial culture enhanced photoheterotrophic growth at low light intensities. The SCP product contained about 65% protein with significant quantities of essential amino acids.

1.2.1.2 Photosynthetic Algal SCP

Among phototrophic organisms several species of algae have been thoroughly investigated as sources of biomass for food or feed. These include: *Chlorella, Scenedesmus, Spirulina, Coelastrum, Dunaliella, Micractinium, Oocystis, Oscillatoria, Chlamydomonas, Euglena, and Ankistrodesmus* [175]. Algae have been used as human food for centuries. *Spirulina platensis* forms part of the diet of natives of the Lake Chad area in Central Africa. A similar algal strain was also used as a food staple by the Aztecs in Mexico before Cortez [175].

Controlled large-scale microalgal culture began in Germany during the 1940's, when experiments were performed to grow certain strains as a source of lipids and protein [175]. During 1951 in Cambridge, Massachusetts, USA, the Arthur D. Little Company experimented with *Chlorella* cultures varying from 2000 to 5000 liters [176]. In 1953, the University of California operated cultures of 10,000 liters, and in 1955, under sponsorship of the Rockefeller Foundation, it constructed and then operated a 2-million-liter pilot plant for 2 years [177]. The work was continued by research groups in the USA, Japan, Taiwan, Israel, France, Italy, Mexico, Peru, Thailand, Bulgaria, Czechoslovakia, USSR, Germany, India and several other countries (see Tables 3 and 6) [26, 46, 175, 178]. In all these studies, artificial ponds (basins) were constructed for large-scale algal cultivation. The only place where algae are mass-cultivated in a semi-natural habitat is in Lake Texcoco in Mexico. Here the Sosa Texcoco Company in cooperation with the Institut Français du Petrole (IFP) harvests one ton of dry *Spirulina geitleri (or maxima)* per day, which is sold as human food ("health" food) mainly to USA, Japan, and Canada [46, 50].

A general flow diagram, showing the main steps of mass cultivation and processing of microalgae, is shown in Fig. 5. According to Becker (1975), three dif-

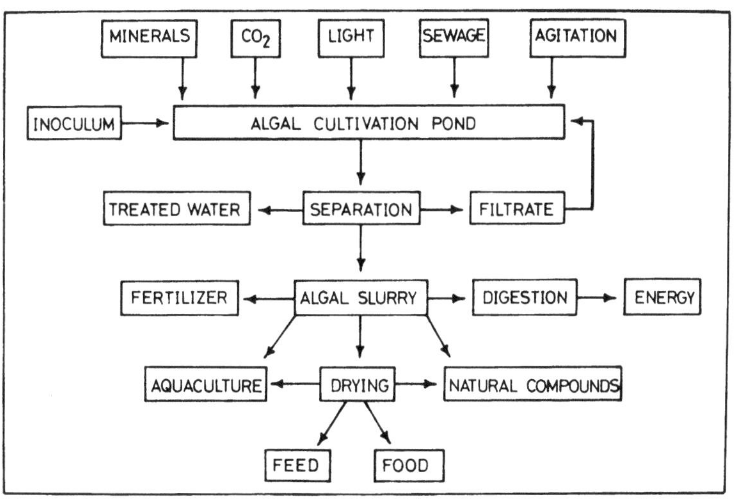

Fig. 5. Flow diagram of the cultivation and processing of microalgae. (Redrawn from Becker [175]). (Reprinted with permission of Process Biochemistry)

ferent systems can be distinguished, depending on the raw materials and the use of the obtained biomass. The first system employs a selected algal strain grown on fresh water, mineral nutrients and additional carbon sources. The SCP is utilized mainly as a food supplement. The second system uses sewage or industrial waste water without the addition of minerals and external carbon. The main importance of this system is in the treatment of waste water and the subsequent use of the biomass as animal feed. In the third system, a food supplement is produced by cultivation of algae in an enclosed system in sunlight or artificial light.

In the first two systems algal cultivation is usually performed in artificial, shallow ponds constructed of various materials such as plastic sheets, concrete, bricks, and glass-fiber laminate. Size varies from a few square meters to several thousand square meters and shape varies from circular basins to long channels [175]. Light is a most important factor in large-scale photosynthetic algal cultivation systems, and therefore lack of cloud cover and minimal diurnal variations in light intensity are required. Growth is limited to areas with suitable climate, as temperatures of about 20 °C for most of the year are required. In open ponds, growth occurs in the top 20 to 30 centimeters, due to limited light penetration through the medium. Therefore, the culture medium has to be continuously or intermittently agitated sufficiently to prevent settling of the algae, to prevent thermal stratification, and to make light and nutrients available uniformly to all the algal cells. This can be achieved by paddle wheels, cage rotors, diffusers, jet pumps, windmills, and even manual agitation (in India) [175].

It can be seen that algal mass cultivation in an open pond system combines characteristics of agriculture (extensive land, water, and fertilizer use, dependence on sunlight and climate, and use of open systems causing contamination by different organisms), and of industrial processes (continuous operation and provision of nutrients at optimal concentrations) for SCP production. Although, in principle, advantages of both agriculture and industrial SCP production could be incorporated into algal mass production, resulting in an economic source of protein, in practice, this technology is still of limited use because of reasons which are stated below [46].

The concentration of biomass obtained, is often limited by the low amount of carbon in the culture medium. The main source of free dissolved CO_2 for algal photosythesis is produced by bacterial respiration of organic compounds. The addition of CO_2 (either in pure form or as waste gas from combustion engines or fermentation processes), and waste organic materials (molasses, cow manure, sewage, industrial wastes) will result in higher photosynthetic productivities [53]. Still, cell densities seldom exceed 1 to 2 g/liter (dry wt) and the productivities obtained are between 10 to 30 $g/m^2 \cdot day$. Although exceptionally high productivity values, such as 95 $g/m^2 \cdot day$, have been reported [179], it is doubtful whether such values can be attained for long periods. As Becker [175] has noted, a mean yield of 20 $g/m^2 \cdot day$ is equivalent to a yield of 70 tons of algal dry matter, or 35 tons of algal protein, per hectare per year. These values are much higher than those obtained with conventional agricultural crops, i.e., wheat (360 kg protein), rice (600 kg protein) and potato (800 kg protein).

The idea of producing algae in conjunction with the treatment of sewage, has attracted much interest mainly from an economical standpoint. By changing the regular algal pond to a system of shallow meandering canals, equipped with mixing and aeration devices, it is possible to obtain an effective algal growth system known as high-rate algal (or oxidation) ponds (HRAP) [26, 46, 180]. The symbiotic process consists of sewage decomposition by bacteria releasing CO_2, whilst the algae, in the presence of solar radiation, use the CO_2 and other nutrients (NH_4, PO_4) to synthesize algal cell material. This photosynthetic reaction, besides producing algae, releases oxygen, which in turn is utilized by the bacteria for sewage decomposition (see Chapter 2 in this monograph for a detailed description of a HRAP system).

The yield of cell biomass, which also expresses the efficiency of the process, is defined as the biomass produced per unit light energy utilized (Y). The photosynthetic efficiency (PE) of a growing algal culture can be determined from the growth yield and the caloric value of the biomass (k), $PE = k \cdot Y$. To obtain maximum photosynthesis the algae have to be grown in light-limited chemostat cultures. Optimal media used are nitrogen and mineral salts media with CO_2 bubbling, described for *Chlorella* 211/8k [167] and *Spirulina platensis* [181]. Using a mixed culture MA003, composed of an algae and three species of heterotrophic bacteria, it was deduced that the maximum practicable storage of total solar energy by algal biomass growth in vitro was 18% [167]. This being about three times the maximum predicted for conventional agriculture. Based on these encouraging results, construction of novel, fully controlled glass or plastic photobiological reactor was suggested for a very large-scale chemical and biomass production process from CO_2 and light energy [183].

Harvesting algae is a particularly difficult problem due to low cell concentration, large volumes of water that must be handled and their small size (10 μm). The harvesting process involves three major steps:

1) initial concentration of the algal suspension,
2) dewatering the resulting slurry, and
3) drying the dewatered algae for storage and handling [177].

Spirulina sp. and several other blue-green algae, such as *Oscillatoria, Microcystis* and *Anabaena*, float to the surface in relatively large clumps and may be skimmed off easily, thus not requiring stages 1 and 2 of the harvesting.

Among the various methods used for concentrating algae from pond effluent are flocculation, sedimentation, filtration, flotation, sand filtration, microstraining, centrifugation, and coagulation; the last two methods are the most frequently applied [46]. In the coagulation system, flocculant particles are formed naturally (i.e., by "autoflocculation" observed with some microalgae as a response to light and temperature changes), or by the addition of reagents such as aluminum sulfate and lime, or through suitable alterations in the chemical and physical constitution of the pond culture [177]. The solid matter ("algal slurry") is then separated from the liquid by flotation techniques. The algal float must next be dewatered, which in most cases is done by centrifugation followed by drying in a drum dryer or spray dryer (in exceptional cases, sun drying or the use of solar heaters is effective). As an alternative to centrifugation, vacuum filtration was also considered, but it was found that direct filtration without filtering aids was not efficient in dewatering the algal float.

1.2.1.3 Chemolithotrophic Bacteria

An alternative method for the production of SCP from carbon dioxide is based on aerobic bacteria, which use molecular hydrogen as electron donor (Table 6). These chemolithotrophic bacteria belong to the genera *Hydrogenomonas* and *Alcaligenes* and are described in numerous review articles [183–189]. The rate of growth of these bacteria with generation times of 3.5 to 7 h surpasses that of the phototrophic microorganisms [190]. Growth is maximal between 28 and 35 °C, at a pH between 6.8 and 7.2. Hydrogen, oxygen and carbon dioxide (as carbon source) have to be supplied, as well as nitrogen, in the form of ammonia, nitrate or urea [34].

The Ajinomoto Company in Japan has been investigating SCP production from carbon dioxide and hydrogen with *Alcaligenes eutrophus* [188, 191, 192]. A maximal yield for H_2 ($Y_{H_2}^{max}$) of 1.9 g cell dry wt/g H_2 was estimated in H_2- and CO_2-limitated cultures, which corresponded well with a value of 1.8 g cell/g H_2 observed for H_2-limited chemostat culture of *Hydrogenomonas* H16.

A Y_{H_2} of 2.3 which exceeded $Y_{H_2}^{max}$ in H_2-limitation was obtained with *A. eutrophus* under O_2-limitation. This probably can be explained by alterations of energetic metabolism [193]. The most efficient utilization of gases is achieved when the supplies of hydrogen, oxygen and carbon dioxide limit the growth simultaneously. The optimum feeding ratio of gases for *A. eutrophus* was estimated as H_2/O_2 ratio of 3 and CO_2/O_2 ratio of 0.4; this ratio is similar to that found for *Hydrogenomonas* H16 [194], strain 9-5 [186, 187] and *Alcaligenes hydrogenophilus* [169]. It was found that under H_2-limiting conditions the cell mass was rich in protein, but the Y_{H_2} was low. Under H_2-limitation, high levels of dissolved oxygen in the medium were toxic for the growth of hydrogen bacteria. Though high Y_{H_2} is expected in O_2-limited chemostat culture, the cell mass obtained in oxygen deficiency is rich in poly-β-hydroxybutyric acid, which has no nutritive value, and therefore a waste of substrates. Similar results were obtained for *Alcaligenes hydrogenophilus* [169, 189].

The main obstacles to SCP production using hydrogen bacteria are those connected with the high cost of gas collection, low solubility of gases in culture media, low productivity (less than 0.5 g cell dry wt/liter · h), need for expensive control devices for the gas mixtures, low concentration of biomass (1–2 g/liter) resulting in the need for recycling and recovery of cells from dilute culture broth, and the problems associated with the very dangerous nature of hydrogen and oxygen handling [40].

1.2.2 Fossil Mass

1.2.2.1 Crude Oil, Gas-Oil, Hydrocarbons

The ability of microorganisms to assimilate hydrocarbons as the sole source of carbon and energy was recognized in 1895 when Miyoshi showed for the first time that the common fungus *Botrytis cinerea* would attack paraffins [195]. This work was followed by Söhngen who, in 1906, isolated the first described methane- and methanol-utilizing bacterium [196] and, in 1913, hydrocarbon-utilizing organisms [197], and by Tausson who studied bacterial oxidation of crude oil in 1928 [198]. However, it was only in 1940, after the specific suggestions by Mogilevskii that sampling oil for hydrocarbon-oxidizing microorganisms could be used as a means for oil prospecting, that an impetus was given to hydrocarbon microbiology [199]. Soon thereafter, tests were performed and, in 1943, the first U.S. patent for oil prospecting was issued, based on this technique [200]. By early 1950, there was sufficient evidence to realize that the ability to utilize hydrocarbons, as carbon and energy sources, was widespread in the microbial world [201]. Since then, thousands of papers and many review articles have appeared on microbial assimilation of hydrocarbons. Much of this information has been summarized for petroleum-degrading bacteria [202–206], yeasts [204–210], and fungi [38, 204–206, 211–213].

Gas-Oil

It was the pioneering work of Alfred Champagnat and his team at British Petroleum in 1963, and the announcement by this company of its intention to manufacture yeast protein from hydrocarbons on a large scale, that stimulated most of the major oil companies (Esso, Shell, Gulf, American, Sun, Mobil) to increase research efforts and to build pilot and production plants for microbiological conversion of petroleum into protein-rich materials (Table 4) [8, 214]. The BP process originated in an attempt to desulfurize petroleum fractions. This had a limited success, but through this work, it was observed that yeasts, particularly *Candida lipolytica*, would grow on petroleum fractions, utilizing the n-alkanes as sole carbon source. Simultaneously, problems concerned with wax crystallization at low temperatures in gas-oil were solved when it was found that yeast fermentations removed wax [8]. However, production of microbial protein from hydrocarbons became the main goal at British Petroleum.

Two processes were developed by British Petroleum. At Lavera, the feedstock was a heavy gas-oil, having a boiling range of 300 to 380 °C and containing 15% C_{15}–C_{30} n-paraffins. The fermentation was run under nonaseptic conditions using *Candida lipolytica* with the recovery of both yeasts for use as animal feed supplement, and dewaxed gas-oil. The yeasts, once produced, had to be exhaustively extracted with organic solvent to remove all possible carcinogens present in the unutilized hydrocarbon fractions (about 90%) of the gas-oil, and flavor taints. At an early stage of the work, it was discovered that the n-paraffin fraction of the gas-oil was being metabolized preferentially by the employed yeast strains [67]. Indeed, it was this finding that led, in 1963, to the development of a second commercial process by BP for yeast production at the Grangemouth refinery in Scotland, based on pure n-paraffins. This was a simpler process as the n-paraffins were almost completely consumed by the yeast throughout the fermentation. Consequently, the post-fermentation stages consisted only of separating and drying the yeast.

Cheaper and easier processes, as compared to yeast fermentation, were developed for dewaxing gas-oil. The n-paraffin fraction obtained (C_{10}–C_{23}) was used as the raw material in the 390-m^3 fermentor at Grangemouth, and later (in 1976), in three 1800-m^3 fermentors at Italproteine at Sarroch Sardinia, constructed by the joint BP-ANIC company. A continuous fermentation was carried out under aseptic conditions, and the intended production capacity was 100,000 ton/year. However, this plant was never allowed to operate commercially, and in 1978, Italprotein was liquidated, BP withdrew from the field and closed its research and development facility for this purpose [8, 215].

Following the pioneering studies of BP on microbial gas-oil and n-paraffin utilization, most of the other major oil firms have taken an active interest in the production of SCP from these raw carbon sources. Semi-commercial plants were built in many countries, namely France, India, Japan, USA, Romania, Switzerland, China, West and East Germany, USSR, Italy, and Czechoslovakia (Table 4). Most of these processes were based on n-paraffins as the raw material, and yeasts as the organism, for SCP production.

The hydrocarbons present in crude oil can be divided into five groups: n-alkanes (or n-paraffins), isoalkanes, alkenes, cycloalkanes, and aromatics [201]. Methane will be discussed in another section of this monograph.

n-Paraffins

n-Alkanes are saturated, straight-chain hydrocarbons, commonly called paraffins. The lowest homologues of the paraffin series, which are liquid at room temperature (C_5 to C_8), are generally toxic to microorganisms. Therefore, the n-paraffins considered as raw carbon sources for SCP production are those having more than 9 carbon atoms [205].

Average n-paraffin content in crude oil is about 5 to 6 percent, obtained by conventional processes. For economic reasons, crude oils containing a high proportion of the required alkanes of given chain length are preferred for separation of n-paraffins. n-Paraffins used as carbon and energy sources for yeast SCP processes are prepared either by urea clathration, or by molecular sieve absorption separation. The latter method is more suitable, yielding n-paraffins (C_{10}–C_{23}) ranging in purity from 97.5 to 99% [216, 217].

Several thousand different yeast, bacterial and fungal strains have been tested for growth on different hydrocarbon substrates. All of the microorganisms which have been found to assimilate liquid n-paraffins can also grow on oxidized carbon sources. In general, their rate of growth on glucose is much higher than on paraffins. Although the capacity to oxidize hydrocarbons is confined to certain genera and species, it is widespread in all of the groups of organisms which have been studied [205].

It is now well known that of the various petroleum hydrocarbons, the n-paraffins of intermediate chain length (C_{10}–C_{19}) are most suitable for microbial utilization. In a comparative study [204], the ability of petroleum-degrading yeasts, fungi, and bacteria to degrade mixed hydrocarbon substrates such as aliphatic, alicyclic, aromatic, and polynuclear aromatic hydrocarbons, was measured. Normal alkanes were found to decrease susceptibility to degradation by bacteria and yeasts, with increasing carbon chain length from 10 to 20 carbon atoms. The following order of decreasing utilization of mixed hydrocarbon substrate in yeast genera was observed: *Sporobolomyces* > *Rhodotorula* > *Candida* > *Hansenula* > *Aureobasidium*. Results obtained for some of the fungi showed that there was little correlation between chain length of normal alkane, and susceptibility to biodegradation. In general, the patterns observed for hydrocarbon utilization were similar for the bacteria, yeasts, and fungi. However, the utilization of even n-paraffins by individual isolates varied significantly [204]. In addition, the individual n-paraffins were not degraded uniformly by a given microorganism [218–220], and this phenomenon has several implications. Firstly, the main factor which influences the yield coefficient in hydrocarbon fermentations is the polyauxic degradation pattern for the hydrocarbon substrate [220]. Secondly, the nonuniform utilization of n-paraffins also effects changes in the physical properties, e. g., the freezing point of the deparaffinated remaining gas-oil [220]. Thirdly, different batches of gas-oil have different compositions of hydrocarbons, effecting nonhomogeneous, nonreproducible fermentation conditions. These effects are important from physiological, industrial and ecological viewpoints.

Most SCP processes using purified n-paraffin fractions employ various yeast strains of the genus *Candida*, for example, strains of *C. lipolytica, C. tropicalis, C. paraffinica, C. pichia, C. guilliermondii, C. maltosa, C. novellus, C. oleophila*, and other unnamed species (Tables 4 and 6). The advantages and disadvantages of these processes are listed in Table 7. The Esso-Nestle partnership developed an n-paraffin SCP process based on the bacterium *Acinetobacter cerificans* [221]. Two other pilot-plants using 50-m^3 fermentors were based on *Pseudomonas* sp. [222–223]. However, all these processes were discontinued.

The four-phase system in the fermentor consists of nutrient medium, paraffin droplets, microbial cells, and air. This system must be dispersed (macromixing) and the parafin droplets must be converted to a form which is utilizable by the biological system (micromixing). (See the book of Wen and Fan on Models for

Table 7. Advantages and disadvantages of an SCP process employing yeasts growing on n-paraffins. (Adapted in part from Dimmling and Seipenbusch [217])

+ Availability of large quantities of substrate	– Dependence on OPEC price of crude oil
+ Availability of substrate at one location	– Costly process for the separation and refining of n-paraffins
+ Amount required for SCP is small from overall supply	– Variable composition of the carbon source
+ High cell yield	– Incomplete utilization of carbon source
+ The cell mass is easy to separate	– Poor solubility of the substrate therefore high power consumption in fermentor
+ Protein plant can be part of larger industrial complex, i.e., expansion of existing facility-no need for new transport systems, source of heating and cooling water etc. can use waste heat and water	– High specific O_2 demand
	– Highly exothermic fermentation
	– Moderate growth rate
	– Lower protein content
	– Additional process step for reducing residual paraffin content of the cell mass
	– Long term effects of product on consumer are unknown
	– FDA regulations harsher than for sugar-based industry
	– Process conditions require greater control than that of sugar industry

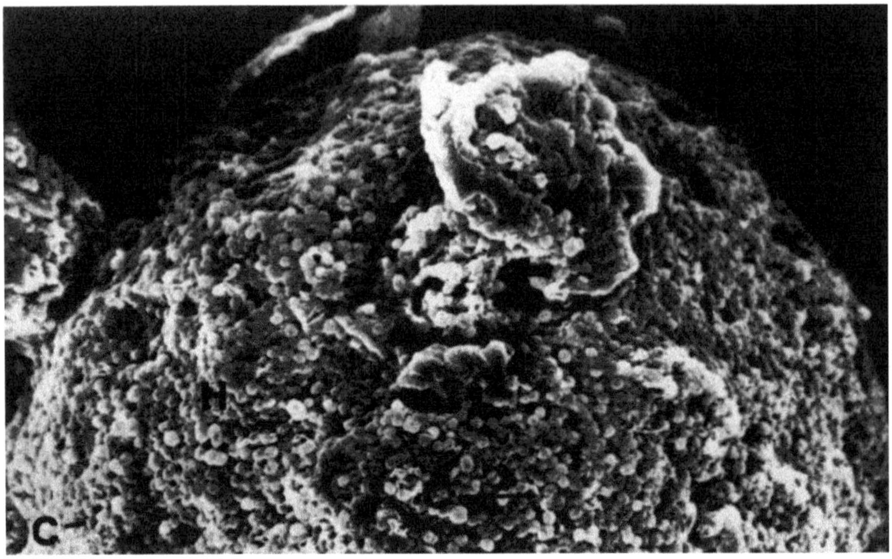

Fig. 6. A scanning electron micrograph of *Acinetobacter* sp. growing on hexadecane. 2000 × . H-Hydrocarbon; C-bacteria. (Reprinted with permission of Springer-Verlag [231])

Flow Systems and Chemical Reactors [224]). Those are the basic operations required for this fermentation, and are connected to the design of a suitable fermentor.

Three possible mechanisms for hydrocarbon transport to microorganisms have been described:

1) via large oil drops,
2) via submicron droplets, and
3) via aqueous dissolution of hydrocarbons [225–229].

In the first, yeasts [229, 230] and bacteria [231] were observed clinging to the oil phase; i. e., transfer of paraffins occurs through direct contact of cells with emulsified droplets of the substrate. Agglomerates between cells and oil droplets were formed, called flocs. In the second and third mechanisms no flocs were observed. In the second, submicroscopical oil droplets (accomodated oil) somehow provide the cells with substrate. The uptake process is not well understood (232–234]; two modes were suggested, namely unmediated transport in yeasts and mediated transport in bacteria [228]. The third mechanism is uptake of dissolved hydrocarbons in aqueous medium [235]; a troublesome physical problem is the limited solubility of hydrocarbons in water. For example, the solubilities of decane, dodecane, and tetradecane in water, at 25 °C, are only 3.1×10^{-7}M, 1.7×10^{-8}M, and 9.8×10^{-10}M, respectively [218]. Therefore, according to the latter mechanism, the growth rate of the microorganism is limited by the transfer rate of the substrate into the solution, and then across the cell wall.

Numerous observations have been made that dispersion of oil in the aqueous broth improves, and the interfacial area between oil drops and water increases as fermentation proceeds. These phenomena are attributed to extracellular microbial products, such as long chain fatty acids [236–238] and various polymeric bioemulsifiers [239], which have already been isolated and identified [225, 240, 241]. Of particular interest is the extracellular product termed "emulsan" which is synthesized by *Acinetobacter calcoaceticus* RAG-1, and produced commercially for various uses [242]. Emulsan (molecular weight, 10^6) is an anionic. D-galactosamine-containing polymer which exhibits substrate specificity with respect to the hydrocarbons it emulsifies, yet is excreted into the growth medium even when RAG-1 is grown on a water-soluble substrate such as ethanol or acetate [243]. Antibodies active against purified emulsan were used in a sensitive enzyme-linked immunosorbent assay to detect free emulsan and emulsan associated with the cell surface [242]. Cells in the early exponential growth phase exhibited relatively large amounts of cell-associated emulsan which decreased rapidly between the midexponential and early stationary phases. This change was accompanied by a rise in the concentration of extracellular free polymer. The cell-associated emulsan, although it resembles the cell-free bioemulsifier antigenically, differs functionally from the isolated polymer in that it cannot emulsify the hydrocarbons. Clearly, more research is needed to clarify this emulsification mechanism.

It is generally agreed that the main pathway for microbial oxidation of alkanes starts at the terminal carbon, and leads to the successive formation of the corresponding primary mono-alcohol, aldehyde and fatty acid [244–247]. The pathway shown in Fig. 7a, represents a hydroxylation reaction, catalyzed by a

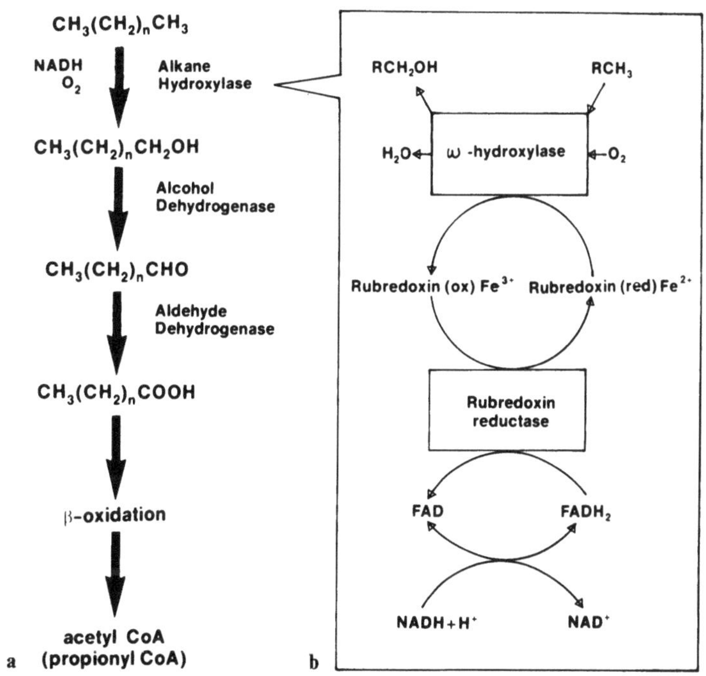

Fig. 7. **a** Alkane oxidation pathway, **b** *Pseudomonas* ω-hydroxylase system

mixed-function oxidase (or monooxygenase), which converts the alkane molecule into a mono-alcohol. This reaction proceeds in the presence of a reductant and molecular oxygen. The n-paraffin (octane) ω-hydroxylating system, purified from *Pseudomonas oleovorans*, consists of three protein components: a flavoprotein reductase, rubredoxin, and a ω-hydroxylase (Fig. 7b) [248–250]. In other organisms cytochrome P450 has been found in place of rubredoxin [205]. In the latter organisms, the alkane hydroxylase is specifically located within microsomes along with cytochrome P450 and NADPH-cytochrome C reductase. Further conversion occurs by alcohol dehydrogenase and aldehyde dehydrogenase, which like the alkane hydroxylase appears to be membrane bound, and by the β-oxidation pathway of fatty acids (Fig. 7a) [240, 251].

The ability of bacteria to grow on n-alkanes of 6 to 10 carbon atoms is made possible by virtue of a transmissible OCT plasmid. Our knowledge of the genetic control of alkane oxidation by alk^+ *Pseudomonas putida* strains is summarized elsewhere [252, 253].

The biochemical pathways by which n-paraffins are assimilated into cell constituents were studied in the yeast *Candida* 107, grown in a continuous culture on C_{13} to C_{16} paraffin molecules [254]. In alkane-grown cells, lipogenesis is repressed, thus carbon flows from the alkane via acetyl CoA, oxaloacetate and pyruvate into pentoses and hexoses in a unidirectional manner. This is due to the strong repression of pyruvate kinase, and the increased activities of phosphoenolpyruvate kinase and fructose-1, 6-diphosphatase enzymes under these conditions.

Different microorganisms undergo various ultrastructural changes specific to, and characteristic of alkane utilization. For example, a complex intracellular membrane system in *Acinetobacter* is induced [231, 255], and in yeast (i.e., *Candida tropicalis*) numerous organelles surrounded with unit membrane envelopes are observed [256]. These changes are accompanied with qualitative and quantitative differences in the lipid composition of the alkane-grown microorganisms, which will be discussed later in this monograph.

Publications describing the fundamentals of n-paraffin-based SCP processes show that all processes which have reached commercial level are very similar in their basic processing steps (Table 4). They differ in microorganism strains and substrates (paraffin fraction) used, fermentor construction and energy balance.

Many engineering problems at each stage of the process diminish the possibility of economic commercialization of n-paraffin-based SCP production [257]. Some of the most important of these problems include poor substrate transfer and solubility, high oxygen demand and heat load, and difficulties in cell separation and purification.

The oxygen requirement in hydrocarbon fermentation is very high since the carbon source contains no molecular oxygen in its structure. The higher oxygen requirement, relative to other fermentation systems, results in the need for an efficient oxygen transfer system. It was found in the Hoechst AG process that the "tubular loop" fermentor with liquid jet propulsion was particularly suitable as an efficient gas-liquid contacting device, as well as a mixing vessel for homogenization and emulsification [65]. Airlift fermentors of various designs were used in the Gulf, Kanegafuchi, IFP, and Liquichimica processes (see the reviews by

Prokop and Sobotka [258] and Faust and Prave [259] on the large numer of fermentor types used in hydrocarbon fermentation). In any case, the high oxygen requirement creates the problem of higher total cost which upsets the economic gain of lower substrate cost.

High oxygen demand yields a proportionally high heat load, effecting, in turn, a high cost of cooling, consequently increasing the cost of the SCP product. Mechanical refrigeration or other costly methods were used for heat removal. Kanazawa [260] described a "vacuum self-evaporated" cooling system for use in the Kanegafuchi yeast SCP process. Data is not available on the suitability of this system for use in production-scale operations.

Commonly-used methods of cell separation are centrifugation and filtration, the choice of which depends upon the size of the microorganism [257]. Two-stage centrifugation for cell separation and washing, followed by spray drying is usually employed [216]. The most serious difficulty in the recovery and purification of cells from fermentation broths arises from the presence of large amounts of unused paraffins. Following fermentation, a processing stage is added in which the cells are held for several hours without any addition of n-paraffins in order to reduce unused hydrocarbon content [261, 262].

Isoalkanes, Alkenes, Cycloalkanes and Aromatics

These compounds have little relevance to SCP production, as they are less easily utilized by microbes than n-paraffins. Branching of the carbon chain (isoalkanes) has been found to decrease metabolic availability in most cases [263]. Hydrocarbons having a double bond (alkenes) are most susceptible to microbial oxidation than their saturated counterparts. However, polyunsaturated olefins are much more resistant to microbial attack than either the alkanes or monounsaturated olefins [201]. Other configurations in hydrocarbon molecules, such as polycyclic structures, yields lower susceptibility to microbial attack.

1.2.2.2 One-Carbon (C_1)-Compounds

One-carbon (C_1)-compounds are defined as reduced compounds containing one or more carbon atoms but containing no carbon-carbon bonds. These include: methane, methanol, methylamine, formaldehyde, formic acid, formamide, carbon monoxide, and other substrates containing more than one carbon atom such as dimethyl ether, dimethylamine, trimethylamine, tetramethylammonium, trimethylamine N-oxide and trimethylsulphonium.

The C_1-compounds which are considered as most suitable feedstocks for SCP production are methane and methanol.

Methane and Methanol

The ability of microorganisms to assimilate non-autotrophically reduced C_1-compounds was first reported in 1906 by Söhngen [196], who isolated a methane-utilizing bacterium, *Bacillus methanicus*. Fifty years later, Dworkin and Foster [264] reisolated Söhngen's strain as a methane- and methanol-utilizing bacterium and renamed it *Pseudomonas methanica*. Since then the microbial utilization of methane and methanol has come under intense study, motivated primarily by the potential of these carbon substrates as feedstocks for SCP production. The reader is referred to the following references for useful reviews on various aspects of SCP production from C_1-compounds [18, 265–277].

Methane, the major constituent of natural gas, would be a preferred raw material for reasons of cost and of wide availability as a refinery product, in many parts of the world. Furthermore, it is advantageous in that it is available in high purity and can be recovered without leaving any residues in the final SCP product. On the other hand, methane suffers both from limited solubility, which results in a mass transfer problem, and from excess heat of combustion, due to its reduced state, which results in a high oxygen demand and excessive cooling requirements. Another limitation is that, due to explosive hazards, operation below 12.1% by volume of oxygen is required. Because of these limitations the commercialization of SCP from methane was not realized.

Methanol is a particularly suitable carbon source for the industrial production of SCP (Table 8). Methanol can be obtained from various sources such as: natural gas, naphtha, heavy fuel oil, coal and cellulose (via synthesis gas). The world capacity for methanol production is about 10 million tons with most commercial processes using natural gas or naphtha. Though it is difficult to forecast future supply rates (from petrochemicals) and prices of methanol, it can be expected that methanol will be available for many years due to the versatility and flexibility of the natural and renewable resources from which it can be produced. Additional reasons for the use of methanol for SCP production are: methanol mixes well with water and has a low boiling point, it is easy to transport and store, and it is produced at a constant composition and has high purity (>99.58%). Methanol is available at a relatively low price, and is partially oxidized so that the oxygen re-

Table 8. Advantages and disadvantages of an SCP process based on bacteria growing on methanol

+ Methanol is obtained from petrochemicals, biomass and other sources	− Lower cell yield than hydrocarbons
+ Methanol of high purity is available	− Higher costs for the maintenance of sterility due to neutral pH
+ Ease of storage and handling	− Methanol is not waste product like methane
+ Complete solubility in water, therefore good substrate transfer in fermentor without undue energy requirements	− Relatively high vapor pressure of methanol at fermentation temperatures
+ Reduced O_2 requirement and heat load relative to hydrocarbons	− Higher nucleic acid content of product requiring extra downstream processing
+ High operating temperature in fermentor possible	
+ High growth rate	
+ High protein content	
+ Large scale production and low price	
+ Restricted use by microorganisms	
+ No problems regarding removal of residual substrate	
+ Less complex design engineering of system than gaseous substrate processes	

quirement and heat load per unit of biomass produced are not prohibitive. Methanol is utilized by a limited number of microorganisms which convert it mainly to biomass, CO_2 and H_2O, and any residual methanol is easily removed from the SCP product. Therefore, it is not surprising that the largest viable commercial plant for SCP production (the ICI's "Pruteen" process) employs methanol as the carbon source for growing the bacterium *Methylophilus methylotrophus*.

Since one-carbon compounds, such as methane and methanol, occur in abundance throughout nature [278], it is not surprising that a wide variety of microorganisms that utilize such compounds, methylotrophic microorganisms, have been isolated (see the book, "The Biochemistry of Methylotrophs" by Anthony [18] for a comprehensive survey of these microorganisms).

The methane-oxidizers (methanotrophs), are primarily responsible for recycling of methane back to carbon dioxide, either directly by respiration, or indirectly, as a result of subsequent degradation of their biomass by aerobic and anaerobic heterotrophs [271, 278].

Methylotrophs also play a significant role in the nitrogen cycle. Many of them oxidize ammonia to nitrite or nitrate and may contribute to the oxidative segment of the nitrogen cycle [279, 280]. There have been a number of reports demonstrating that methane-utilizers fix atmospheric nitrogen [279, 281, 282], and that methanol-utilizers can be used efficiently in biological denitrification processes [283]. Thus, methylotrophic processes greatly contribute to the ecological balance of numerous niches and to the world at large.

Whittenbury et al. [281] isolated over 100 strains of methane-utilizing bacteria from various sources (see Table 9 for a selected list). The bacteria were found to be strict aerobes, and were only able to grow on methane, methanol, or dimethylether, but not on multicarbon compounds (obligate methanotrophs). They are recognized in that they are Gram-negative rods, vibrios or cocci, catalase-positive and have typical cytochromes, a, b, and c. To some degree, they all are sensitive to normal oxygen tensions in air. If N_2 fixation is induced, they become extremely oxygen sensitive. All use ammonia and most use nitrate and nitrite as the nitrogen source. Some can utilize urea, amino acids and yeast extract and all reduce nitrate to nitrite. All methanotrophs can oxidize ammonia, methane, methanol, carbon

Table 9. Selected methane-utilizing bacteria

Methylomonas methanica	*Methylococcus capsulatus*	*Methylococcus vinelandii*
Methylomonas carbonatophila	*Methylococcus capsulatus strain Bath*	*Methylosinus sporium*
Methylomonas rubrum		*Methylosinus trichosporium*
Methylomonas rosaceus	*Methylococcus ucrainicus*	*Methylosinus trichosporium TG*
Methylomonas agile	*Methylococcus fulvus*	*Methylocystis parvus*
Methylomonas albus	*Methylococcus thermophilus*	*Methylobacterium organophilum*
Methylomonas streptobacterium	*Methylococcus albus*	*Mycobacterium methanicum*
Methylomonas methanooxidans	*Methylococcus minimus*	*Mycobacterium cuneatum*
Pseudomonas methanitrificans	*Methylococcus luteus*	*Nocardia rhodochrous*
Pseudomonas strain L-8	*Methylococcus bovis*	*Nocardia ucrainica*
Pseudomonas strain L-47	*Methylococcus chroococcus*	*Rhodopseudomonas gelatinosa*
Pseudomonas strain L-49	*Methylococcus whittenburii*	*Strain TM-10*

Based on Foo [284] and Romanovskaya et al. [285]

monoxide, dimethyl ether, propane, ether, ethanol, propanol, butanol, formaldehyde and formate, and other compounds.

The methanotrophs possess a complex internal arrangement of paired membranes and form a differentiating resting body (exospore or cyst). The majority of methane-utilizers are mesophiles, though some are able to grow at 50 °C (thermotolerant) but their optimum growth temperatures are near 37 °C. Two thermophilic methane-utilizing bacteria, *Methylococcus thermophilus* and strain H-2, have an optimum temperature range for growth of 50 to 55 °C [286, 287].

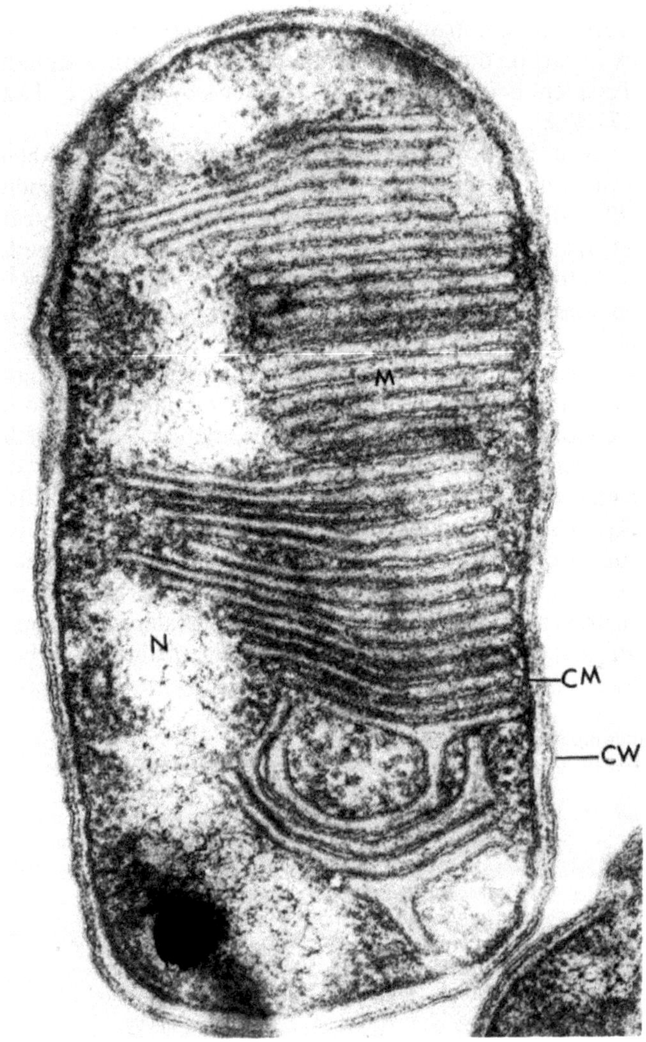

Fig. 8. Section of a methane-grown *Methylomonas* sp. Cell wall (CW), cytoplasmic membrane (CM), internal membrane stacks (M), and nuclear material (N) are shown at 123,000 ×. (Reprinted with permission of De Boer and Hazeu [289])

Whittenbury et al. [281] and Lawrence and Quayle [288] classified the obligate methane-utilizing bacteria into two groups, Type I and Type II, according to several characteristics. Type I methanotrophs have the following characteristics: formation of intracytoplasmic membranes arranged in bundles of vesicular discs (Fig. 8), formation of cysts as resting stages, assimilation of methane and methanol via the ribulose monophosphate (RMP) pathway (see below), the presence of an incomplete tricarboxylic acid (TCA) cycle, and the predominant biosynthesis of C_{16} fatty acids in the membranes. In addition, Type I methanotrophs use glutamate dehydrogenase (GDH) when grown in high ammonia and their glutamate synthase (GOGAT) is NADPH-dependent [290–292].

By comparison, Type II methanotrophs have the following characteristics: formation of intracytoplasmic paired membranes around cell periphery (Fig. 9), formation of exospores or lipid cysts as resting stages, assimilation of C_1-compounds via the serine pathway (see below), the presence of a complete TCA cycle, the predominant biosynthesis of C_{18} fatty acids in the membranes, the absence of GDH activity, and the possession of a NADH-dependent GOGAT activity.

It was only in 1974 that Hanson and his colleagues first described a facultative methanotrophic bacterium, *Methylobacterium organophilum* [294], and in 1979

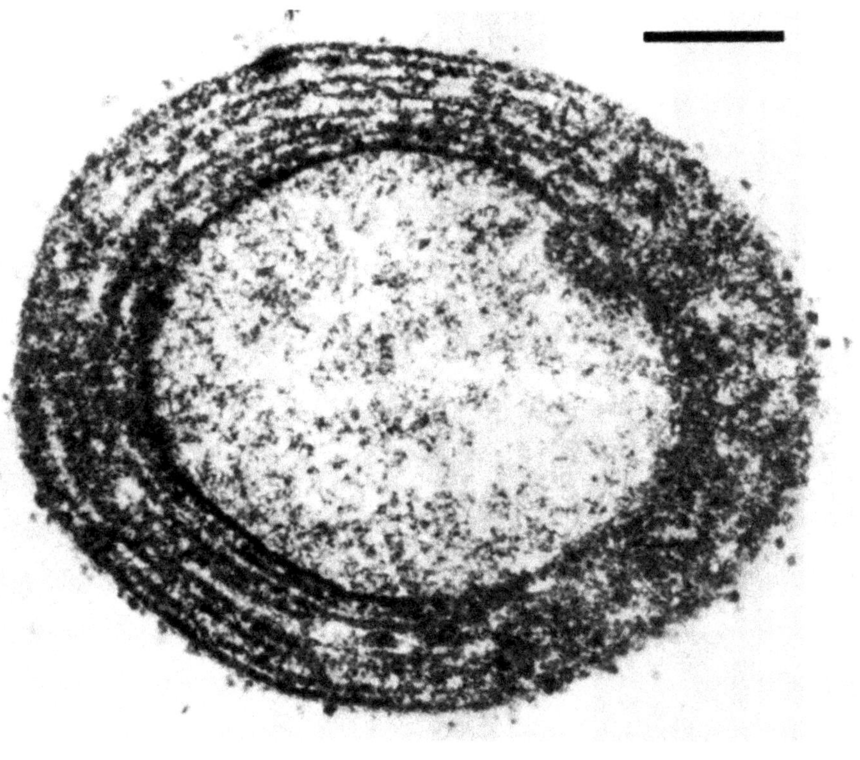

Fig. 9. Section of a methane-grown *Methylobacterium organophilum strain* XX. The bar represents 0.2 μm. (Reprinted with permission of Patt et al. [293])

Table 10. Methylotrophic microorganisms able to grow on methanol but not on methane. S-obligate methylotrophic bacteria able to grow on methanol and methylamine., C-pink facultive methylotrophic bacteria., P-methylotrophic yeasts isolated on methanol

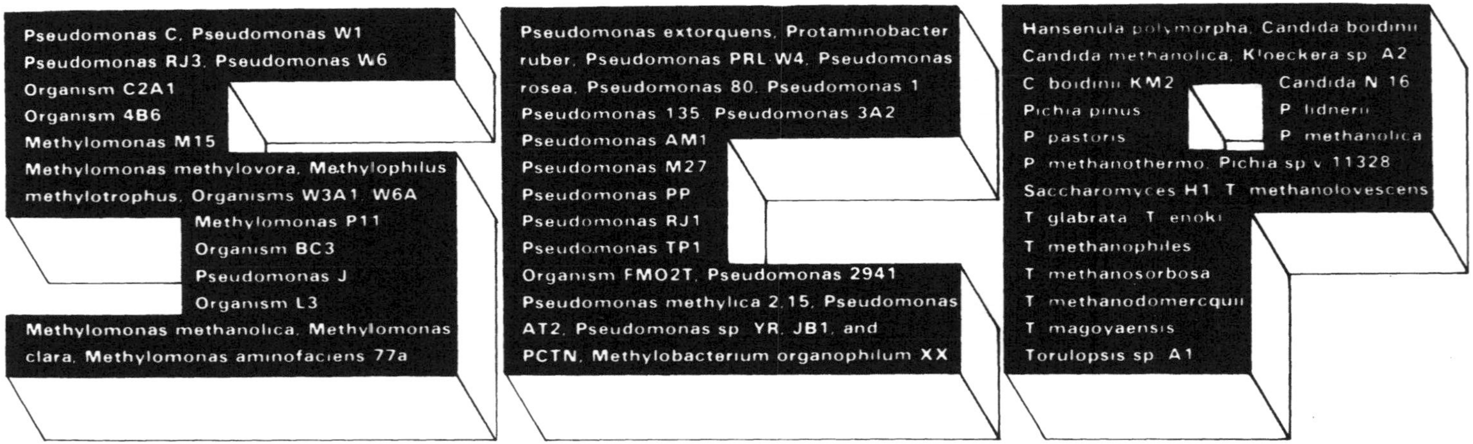

that they first described five facultative methanotrophic yeasts [295]. Several algae and fungi are thought to have a limited capacity to utilize methane; however, the evidence is inconclusive [271].

A large number of bacteria able to grow on methanol, but not on methane, have been isolated and characterized (Table 10) [18, 270, 278, 296]. Some of these bacteria are obligate methylotrophs, able to grow only on methanol and methylated amines and utilize these compounds via the RMP pathway (see below). The majority of the bacteria isolated from enrichments using methanol and methylamine as substrates, also grow on other C_1 and non-C_1-carbon substrates, and hence these bacteria are named facultative methylotrophs. The majority of these isolates are pink-pigmented, Gram-negative bacteria, belonging to *Pseudomonas, Vibrio, Protaminobacter, Flavobacterium, Achromobacter* and *Hyphomicrobium* (Table 10) [297]. A few Gram-positive methanol-assimilating bacteria belonging to *Arthrobacter, Bacillus* and *Corynebacterium* genera have been isolated [297]. All the facultative methylotrophic bacteria assimilate C_1-compounds via the serine pathway (see below).

Ogata et al. [298] first described the isolation of numerous methanol-utilizing yeasts from natural sources, and subsequently, many more strains were isolated (Table 10) [299]. Lee and Komagata [300] divided the methanol-utilizing yeasts into four groups: *Candida boidinii* group, methanol-assimilating *Hansenula* group, *Pichia cellobiosa* group, and the *Hansenula capsulata* group. One mycelial fungus, *Trichoderma lignorium* [301], and one *Actinomycete*, *Streptomyces* sp. [239, 302], are capable of utilizing methanol as a sole carbon source.

Carbon Metabolism. All methylotrophs are able to synthesize cell material entirely from one-carbon compounds. As with CO_2-fixing autotrophic organisms

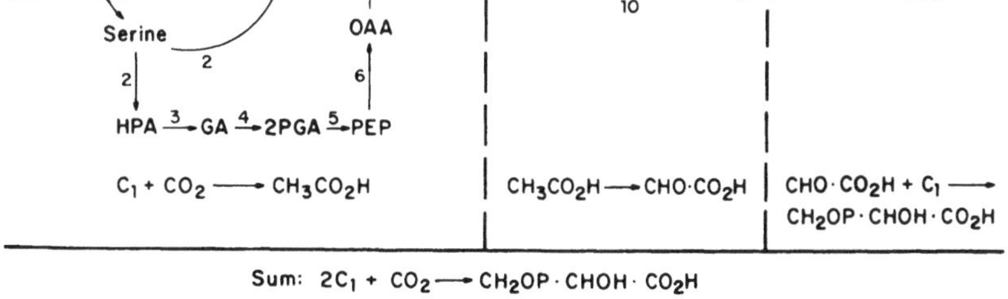

Fig. 10. The isocitrate lyase^{+}-serine pathway. The abbreviations used are: PEP, phosphoenolpyruvic acid; PGA, 3-phosphoglyceric acid; HPA, hydroxypyruvic acid; OAA, oxaloacetic acid; GA, glyceric acid; 2PGA, 2-phosphoglyceric acid. 1. Serine transhydroxymethylase. 2. Serine-glyoxylate aminotransferase. 3. Hydroxypyruvate reductase. 4. Glycerate kinase. 5. Enolase. 6. PEP-Carboxylase. 7. Malate dehydrogenase. 8. Malate thiokinase. 9. Malyl-CoA-lyase. 10. Isocitrate lyase. (Reproduced with permission from Biochemical Society Transactions [273])

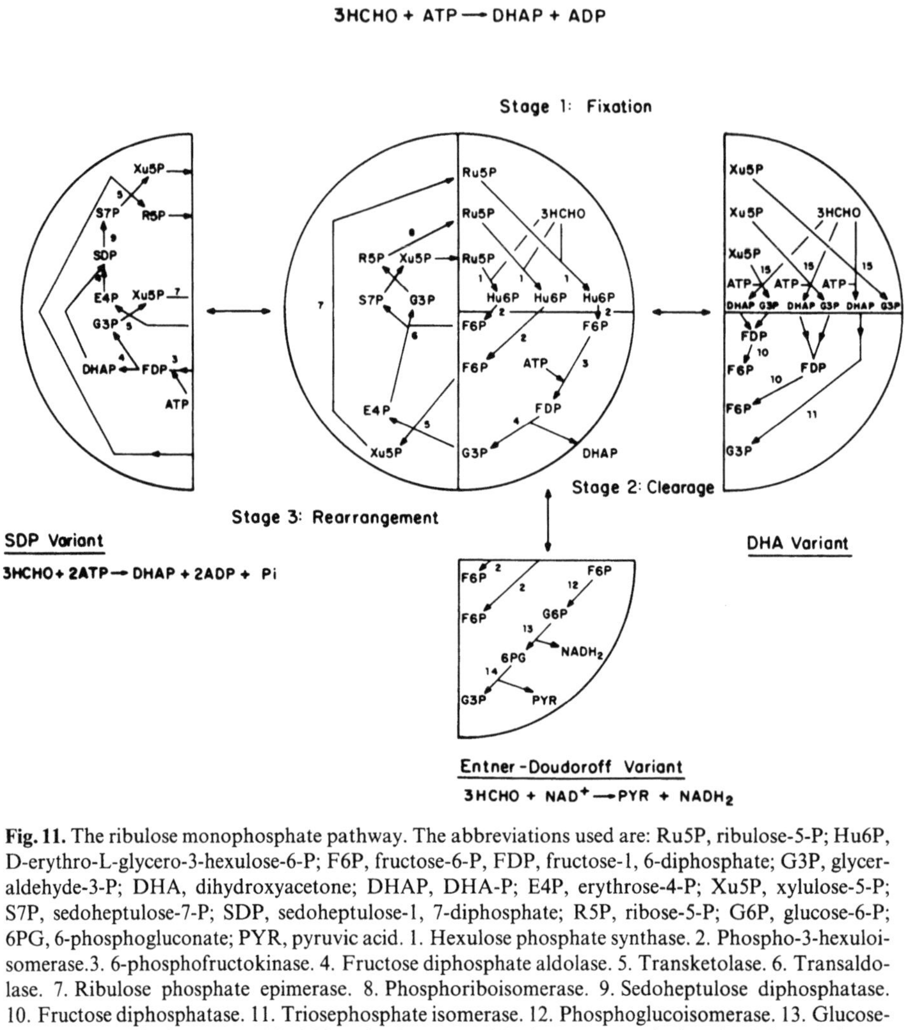

Fig. 11. The ribulose monophosphate pathway. The abbreviations used are: Ru5P, ribulose-5-P; Hu6P, D-erythro-L-glycero-3-hexulose-6-P; F6P, fructose-6-P, FDP, fructose-1, 6-diphosphate; G3P, glyceraldehyde-3-P; DHA, dihydroxyacetone; DHAP, DHA-P; E4P, erythrose-4-P; Xu5P, xylulose-5-P; S7P, sedoheptulose-7-P; SDP, sedoheptulose-1, 7-diphosphate; R5P, ribose-5-P; G6P, glucose-6-P; 6PG, 6-phosphogluconate; PYR, pyruvic acid. 1. Hexulose phosphate synthase. 2. Phospho-3-hexuloisomerase.3. 6-phosphofructokinase. 4. Fructose diphosphate aldolase. 5. Transketolase. 6. Transaldolase. 7. Ribulose phosphate epimerase. 8. Phosphoriboisomerase. 9. Sedoheptulose diphosphatase. 10. Fructose diphosphatase. 11. Triosephosphate isomerase. 12. Phosphoglucoisomerase. 13. Glucose-6-phosphate dehydrogenase. 14. 6-Phosphogluconate dehydrogenase, and phospho-2-keto-3-deoxygluconate aldolase. 15. Transketolase and triokinase. (Reproduced with permission from Annual Reviews of Microbiology, Vol. 33, 1979 by Annual Reviews, Inc.) [270]

(the Calvin pathway; Fig. 4), a three or four-carbon skeleton, such as pyruvate or succinate, is synthesized from C_1-units. Subsequent to intermediate formation, the biochemical pathways by which other cell constituents are synthesized are similar to those found in heterotrophic microorganisms [271–273].

Two assimilation pathways for reduced C_1-compounds exist in methylotrophs: the serine pathway (Fig. 10) [296, 302], and the ribulose monophosphate (RMP) pathway (Fig. 11) [303, 304].

The Calvin, serine, and RMP pathways, which effect assimilation of one-carbon compounds, differ in the nature of the acceptor molecule of the entering C_1-units. The acceptor molecule in the Calvin cycle is ribulose diphosphate, in the serine pathway, glycine and phosphoenolpyruvate, and in the RMP pathway, ribulose monophosphate.

A further difference between the three assimilation pathways is the oxidation level of the entering C_1-units. Carbon dioxide alone is incorporated to form a C_3-compound by the Calvin cycle. Formaldehyde and CO_2 are incorporated by the serine pathway, whereas only formaldehyde is fixed via the RMP pathway. As will be shown later, the oxidation level of the fixed C_1-substrate is important for SCP production as it determines parts of the overall efficiencies of the different assimilation pathways and hence, affects the cellular yields obtained.

A few microorganisms use the Calvin cycle during growth on reduced C_1-compounds. Examples include: aerobic-autotrophic metabolism of formic acid by *Pseudomonas oxalaticus* [305, 306] and *Thiobacillus novellus* [266], aerobic growth on methanol by *Paracoccus denitrificans* [307–309], *Achromobacter, Pseudomonas,* and *Mycobacterium* [310], and photosynthetic bacteria [311, 312]. Because the definition of methylotrophs used here excludes organisms which grow autotrophically on reduced C_1-compounds, the above bacteria utilizing the Calvin pathway will not be discussed further.

The Serine Pathway. The serine pathway (Fig. 10) was first suggested in 1961 by Large et al. [302] as a result of short-term isotope incubation studies, using the facultative methanol-grown *Pseudomonas* AM-1 and *Hyphomicrobium vulgare* (the sum of the pathway is: 2 formaldehyde$+CO_2 \rightarrow$2-phosphoglyceric acid). It was subsequently observed that all facultative methylotrophs and Type II obligate methane-utilizers possess high levels of serine pathway enzymes. The serine pathway is based on the condensation of C_1-units (formaldehyde and CO_2) with C_2- and C_3-compounds. It accomplishes the biosynthesis of oxaloacetate from one molecule of glycine, one equivalent of formaldehyde in the form of hydroxymethyltetrahydrofolate, and one molecule of carbon dioxide. The mechanism by which the glycine skeleton is synthesized in net fashion from C_1-units was suggested by Bellion and Hersh [313]. They found that certain serine pathway organisms synthesize both high levels of isocitrate lyase (icl$^+$-variant), and an ATP and CoA-dependent malate-cleavage system. The complex sequence of reactions participating in the icl$^+$-variant of the serine pathway is shown in Fig. 10. In other serine-pathway organisms, which do not possess malate thiokinase and isocitrate lyase (the icl$^-$-variant) [314], the mechanism of glycine regeneration is not known.

The Ribulose Monophosphate (RMP) Pathway (Quayle Cycle). As a result of studies with the methane-utilizers *Pseudomonas methanica (Methylomonas methanica)* and *Methylococcus capsulatus*, the ribulose monophosphate (RMP) pathway of formaldehyde fixation (Fig. 11) was proposed by Quayle and coworkers [303, 315, 316] (the sum of this pathway is: 3 formaldehyde$+$ATP$\rightarrow$dihydroxyacetone phosphate$+$ADP). Subsequent work showed that obligate methanotrophs and methanol-utilizing bacteria possess this pathway. The RMP pathway is conveniently divided into three stages (Fig. 11). Stage 1 involves the aldol condensation of three molecules of formaldehyde with three molecules of ribulose-5-phosphate to yield three molecules of an unusual 3-oxohexulose 6-

phosphate (referred in Fig. 11 as erythro-L-glycero-3-hexulose 6-phosphate), and further on, three molecules of fructose 6-phosphate. These reactions are common to all methylotrophic bacteria using this pathway [270]. Stage 2 involves cleavage of the hexose phosphate to triose phosphate (C_3-compounds), either by fructose diphosphate cleavage as in Embden-Meyerhof pathway, or the Entner-Doudoroff pathway (Fig. 11). Stage 3 of the RMP pathway involves the regeneration of three molecules of ribulose-5-phosphate through mechanisms involving either transaldolase (regeneration from fructose-6-phosphate and glyceraldehyde-3-phosphate) or sedoheptulose-1,7-diphosphate and fructose diphosphate aldolase (Fig. 11). The two routes of cleavage can be coupled to two methods of rearrangement, thus forming four possible variants of the RMP pathway.

Originally, it was thought that a particular C_1-utilizing microorganism possessed either the Calvin cycle, the serine pathway, or the RMP pathway to the exclusion of the other two [296, 317]. However, evidence has accumulated that several microorganisms may contain more than one biochemical pathway for C_1-assimilation, either constitutive or induced under different growth conditions. For example, the methane-utilizer *Methylococcus capsulatus* (Bath) contains key enzymes of the Calvin and the RMP pathway [318], and the methanol-utilizer *Streptomyces* sp. [319] or the methane-grown *Methylococcus capsulatus* (Texas) and *Pseudomonas methanica* [314], contain enzymes of the RMP and the serine pathway.

The Pentose Phosphate Cycle of Formaldehyde Fixation in Yeasts. The incorporation of ^{14}C-labeled C_1-compounds into phosphate esters of fructose and glucose in methanol-grown yeasts was first reported by Fujii and his colleagues [320, 321]. The absence of both hydroxypyruvate reductase (a key enzyme in the serine pathway) and ribulose diphosphate carboxylase (a key enzyme in the Calvin cycle), together with the presence of the above labeling pattern was used as evidence for the operation of an RMP pathway in yeasts. However, the specific activity of hexulose phosphate synthase – a key bacterial enzyme in the RMP pathway – was too low to account for growth of yeasts on methanol and, unlike the bacterial enzyme, the yeast enzyme was stimulated by ATP [320, 321]. Furthermore, Kato et al. [322] could not detect the RMP-pathway enzyme phosphohexulose isomerase in cell-free extracts of *Kloeckera* sp. 2201. This dilemma was solved by Quayle's group when they demonstrated that condensation of formaldehyde and xylulose monophosphate, catalyzed by dihydroxyacetone (DHA) synthase, yielded dihydroxyacetone phosphate (DHAP) and glyceraldehyde-3 phosphate (Fig. 11) ([273], see also [323]). Recently, DHA synthase was purified from methanol-grown *Candida boidinii* 2201 and found to be a new type of transketolase with respect to its specificity for ketol acceptors [324].

Energy Metabolism in Methylotrophs. Methylotrophs oxidize C_1-growth substrates to CO_2 via special oxidation pathways, shown in Fig. 12. From these oxidative reactions, ATP, is derived, and used subsequently for biosynthesis of cell constituents. Formaldehyde occupies a central position in the metabolism of methane and methanol because at the formaldehyde oxidation level, carbon is both assimilated into biomass and oxidized to CO_2 to provide energy for growth.

The Direct Oxidation Pathway. The oxidation of C_1-compounds in most methylotrophs proceeds via a series of reactions leading from methane to methanol to formaldehyde to formate and then to CO_2 – the direct oxidation pathway (Fig. 12) [296]. Evidence for operation of the direct oxidation pathway in methylotrophs, as well as a detailed description of characteristics of the enzymes involved, is given by Anthony [18].

An unusal phenomenon was observed in yeasts, where growth on methanol was accompanied by a pronounced increase in the number of microbodies, known as peroxisomes [325–332]. These organelles contain high levels of flavin-dependent alcohol oxidase [283–299, 330], which catalyzes the reaction:

$$CH_3OH + O_2 \rightarrow HCHO + H_2O_2.$$

The bulk of formaldehyde formed in the peroxisomes is exported to the cytoplasm where it is oxidized to CO_2 by NAD^+-dependent dehydrogenase. Under certain conditions, a portion of the formaldehyde is oxidized to formate in the peroxisomes either by methanol oxidase itself [333], or by the reaction of H_2O_2 and catalase (an enzyme present in methanol induced peroxisomes) [327, 334, 335]. Since the first step in methanol oxidation in yeasts involves a flavoprotein oxidase which is not coupled to ATP synthesis, it might be expected that the energy gained from methanol oxidation by yeasts is lower than that of bacteria [18]. Practical observations have verified this hypothesis (see later in Table 19).

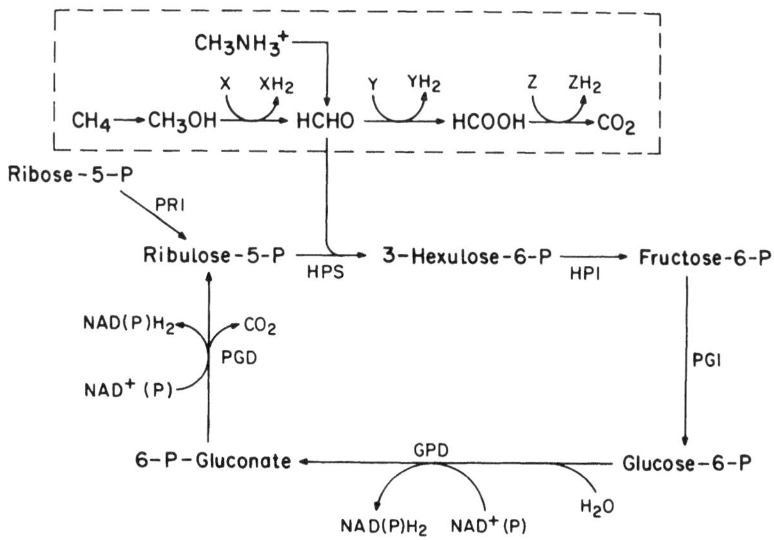

Fig. 12. Pathways for direct and cyclic oxidation of reduced C_1-compounds to CO_2. The abbreviations used are: GPD, glucose-6-phosphate dehydrogenase; HPS, hexulose phosphate synthase; PGD, 6-phosphogluconate dehydrogenase; PGI, phosphoglucoisomerase; HPI, hexulose isomerase; PRI, phosphoriboisomerase; X, Y, Z, represent electron acceptors. Broken line – the direct oxidation pathway

The Dissimilatory Ribulose Monophosphate Pathway (The Cyclic Pathway). Instead of obtaining energy by oxidizing methanol via the direct oxidation pathway by methanol, formaldehyde and formate dehydrogenases, bacteria utilizing the RMP assimilatory pathway have an alternative cyclic oxidation pathway, which branches at the level of formaldehyde (Fig. 12) [304, 336, 337]. It should be noted that the only additional enzyme necessary for operation of such a dissimilatory cycle in an RMP-pathway bacterium using Entner-Doudoroff cleavage (Fig. 11) is 6-phosphogluconate dehydrogenase, which has indeed been detected in many methylotrophs which use the RMP pathway [267, 338]. Evidence for the operation of this oxidation cycle in RMP-pathway methylotrophs has been obtained [339, 340].

Both in the direct oxidation pathway (via formic acid) and the cyclic oxidation pathway, two moles of reduced pyridine nucleotides and one mole of CO_2 are formed for each mole of formaldehyde oxidized (Fig. 12). However, the cyclic oxidation pathway yields both NADPH and NADH, whereas in most cases, the direct oxidation pathway generates only NADH [339], with the exception of an $NADP^+$-linked formaldehyde dehydrogenase described in *Methylococcus capsulatus* (Bath) [341]. There is evidence that both the cyclic and the direct oxidation pathways operate simultaneously in RMP-pathway bacteria, e. g., the methanol-grown *Pseudomonas C* [339]. This raises the question of the mechanism(s) which control the flux of carbon from methanol via the different biochemical pathways, leading to production of either cell biomass or CO_2. Only preliminary studies have dealt with this question [277, 342, 343]. Clearly, more experimental work is needed to gain knowledge to allow the alteration of the carbon flux in these organisms so as to increase the efficiency of conversion of substrate carbon to cell carbon.

The Tricarboxylic Acid (TCA) Cycle. Glucose-6-phosphate is an important intermediate in both the assimilatory and dissimilatory RMP pathways in methylotrophs. Glucose-6-phosphate may be oxidized to CO_2 via the Embden-Meyerhof pathway together with the tricarboxylic acid (TCA) cycle. However, it was found that methylotrophic bacteria which operate the RMP pathway possess an incomplete TCA cycle, i. e., the activity of a key enzyme in the TCA cycle, α-ketoglutarate dehydrogenase, could not be detected in cell extracts [270, 344]. Unlike RMP pathway bacteria, methylotrophs using the serine pathway possess a complete TCA cycle. The role of the TCA cycle in methylotrophs has not been fully elucidated and more detailed knowledge in this area is required.

Several process-research programs were carried out on bacteria growing on methane as a sole carbon source [269]. The process-research work performed at Shell Research Limited, at Sittingbourne, Kent, England, using a mixed culture of bacteria has been well documented [345]. However, due to the limitations associated with the use of methane (see earlier in this section), SCP manufacturing processes from natural gas were not realized, and several companies turned to methanol as a raw material. Single cell protein manufacturing processes from methanol, using bacteria or yeasts, were described by Hamer [269] and Solomons [8] (Tables 4 and 6). A detailed description of the methanol-based ICI's process, the largest operating SCP process, is given in Chap. 2 of this monograph.

Other C_1-compounds

Formaldehyde and formic acid are intermediary metabolites in the oxidation pathway of methane to CO_2 and are present in crude methanol. Feasibility tests for use of these compounds as raw materials for SCP production have been performed [268]. Both formaldehyde and formate can serve as carbon and energy sources for facultative methylotrophs which follow the serine pathway [268, 346, 347]. Two soil fungi, *Gliocladium deliquescens* and *Paecilomyces variotii*, were able to grow on 0.1 to 0.2% formaldehyde or 0.5% sodium formate as the sole carbon sources [278]. For those methylotrophs which utilize the RMP pathway, formaldehyde and formic acid are thought to serve only as energy sources [343, 348, 349].

1.2.2.3 Ethanol

Ethanol may well be one of the most attractive of all substrates for SCP production because it has some very distinct advantages compared to other substrates (these are summarized in Table 11). Ethanol's attraction as a substrate comes mainly from its availability as a petrochemical feedstock and as a major product derived from grain and other renewable resources. Furthermore, the standard homogeneity, purity, nontoxicity, and quality of substrate, guarantees production stability and acceptance of the ethanol-based SCP product for human consumption.

The use of ethanol has a substantial cost disadvantage, i.e., the order of 44–55 ¢/kg compared to 13–20 ¢/kg for glucose in 1981 [8, 60]. It has a relatively high volatility in dilute solution. Therefore, to avoid appreciable evaporative loss of ethanol it must be kept at a relatively low concentration in the fermentation broth. This can be accomplished through utilizing a continuous or fed-batch ("Zulauf") process [51, 60, 350].

In 1978, the world capacity for the production of synthetic ethanol from petroleum-derived ethylene was about 2 million ton/year and that of Europe was 0.6 million ton/year [217]. During recent years, the energy supply situation has rekindled interest in ethanol as a fuel. In several countries, Brazil in particular, ethanol is produced by fermentation on a very large scale for motor fuel (gasohol) use.

Table 11. Ethanol as an SCP substrate – advantages and disadvantages. (Based on Humphrey [60])

+ May be major chemical derived from renewable resources, coal, and petroleum; not necessarily dependent on one specific raw material	– Relatively high substrate cost
+ Available as a very pure material	– Readily utilized by many species of microorganisms as a carbon source, therefore it can be contaminated
+ Acceptable as a food ingredient	– High relative volatility therefore must be used in dilute solutions; not cost effective to use in batch mode; fed-batch operation is suitable
+ Completely water miscible	
+ Not markedly inhibitory to most microorganisms	
+ Lower O_2 demand and heat release than other hydrocarbon substrates	– Higher O_2 demand and heat release than carbohydrate substrates

Thus, in certain developing countries which have no fossil raw materials of their own but have plentiful supplies of carbohydrate sources, large quantities of ethanol can be produced, so that, apart from being used for fuel and in the chemical industry, sufficient quantities can be made available for SCP production. However, these processes are still subsidized by the governments of the respective countries.

Many microorganisms have been reported to utilize ethanol as a sole source for carbon and energy.

Bacteria belonging to the following genera are the most common ethanol-utilizers: *Acetobacter, Acinetobacter, Arthrobacter, Bacillus, Brevibacterium, Corynebacterium, Hyphomicrobium, Nocardia,* and *Pseudomonas.*

Ethanol-utilizing yeasts belong to the genera: *Candida, Debaromyces, Endomycopsis, Hansenula, Trichosporon, Lentinus, Pichia, Rhodotorula, Saccharomyces,* and *Cryptococcus* [51]. Recently, 285 strains of ethanol-assimilating yeasts, isolated from natural and industrial sources, were studied for their physiological and biosynthetic properties [139]. The strains belonged to the genera *Candida, Saccharomyces, Pichia, Trichosporon, Rhodotorula* and *Cryptococcus, Candida kruzei, Candida sorbosa, Candida guiliermondii,* and *Trichosporon cutaneum* produced the highest yields of biomass from ethanol.

The stoichiometric relationship for yeast metabolizing ethanol is given by the equation:

$$8\,CH_3CH_2OH + 9\,O_2 + 2\,NH_3 \rightarrow 2\,C_7H_{11}NO_3 + 2\,CO_2 + 16\,H_2O \quad (1)$$

resulting in a maximal yield coefficient of 0.82 g cell dry wt/g ethanol utilized [60]. Ethanol is first oxidized to acetaldeyhde followed by a second oxidation to acetate (Fig. 13). Thus, two moles of NADH are produced for each mole of ethanol oxidized to acetate. Both acetaldehyde and acetate can accumulate in the broth when ethanol is in excess, thus inhibiting growth, explaining in part, the ethanol toxicity towards microorganisms [51, 351–353]. Another mechanism that underlies the inhibition by ethanol of growth of *Saccharomyces cerevisiae* is the non-competitive inhibition of nutrient transport across the plasma membrane [354–356].

The acetate produced from ethanol oxidation can be accumulated as an extracellular product, or converted into acetyl CoA, depending on the ethanol concentration in the medium and the metabolic activity of the organism. Acetyl CoA enters the TCA cycle through malate or citrate, depending upon the energy status of microorganism. The necessary requirement for the utilization of ethanol (a C_2-compound) by microorganisms is the presence of a gluconeogenic system, which enables the microorganism to synthesize C_3-compounds from C_2-units [60, 352, 357]. This is accomplished by the operation of the TCA cycle together with glyoxylate shunt which provide both energy and cell constituents (Fig. 13).

Numerous organizations and industrial companies in various countries have been actively looking at ethanol as a fermentation substrate for SCP production (Tables 4 and 6). One of the few SCP processes actually in operation, with its product "Torutein" being sold, is that of Pure Culture Products (American Oil

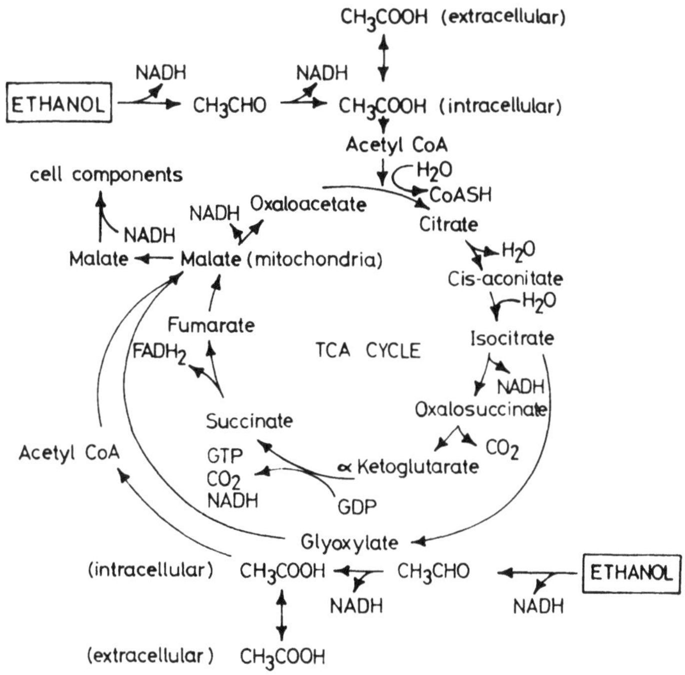

Fig. 13. Biochemical pathways of ethanol utilization. (Reprinted with permission from Humphrey [60], Pergamon Press Ltd.)

Company). In this process *Candida utilis* is grown on food-grade ethanol to obtain a product which is used as a nutritional supplement and flavor enhancer for processed foods [8, 51, 358].

The Mitsubishi Petroleum Company in Japan has been developing strains of *Candida, C. ethanothermophilum* and *C. acidothermophilum*, which can grow on ethanol at temperatures in the range of 40 °C and at low pH [359–361]. A 100 ton/year pilot plant is presently operating with a 100,000 ton/year plant expected to be the minimum-sized commercial plant to be constructed.

A unique process for growing a strain of *Hansenula anomela* on ethanol was developed by the Madrid Institute of Industrial Fermentation [362, 363]. The yeast was grown at a relatively low pH (3.5) which lowered the chances of contamination and on 95% pure oxygen, which prevented foam formation and provided a very high productivity (45 $kg/m^3 \cdot h$; biomass yield of 0.75 g cell dry wt/g ethanol), so that only a relatively small fermentor (20,000 liter) was required to provide a high amount of SCP product.

In the Exxon-Nestle project, a large number of both yeast and bacterial strains were isolated, and considerable process development was undertaken [51]. Greatest attention was concentrated on the use of strains of the bacterium *Acinetobacter calcoaceticus*; the SCP product obtained was planned for food use rather than animal feed.

1.2.3 Renewable Resources as Carbon Substrates for SCP Production

Various types of renewable raw materials, have been used as important industrial feedstocks in the organic chemical and fermentation industries. Although fossil-derived materials have predominated during the last 30 years, and will continue to be the most important organic feedstock for many years to come (at least for the chemical industry), biomass is again receiving serious attention. The competition between biomass and fossil mass is not just a matter of renewable versus non-renewable resource; biomass is also adaptable, cheap, ubiquitous, available and, in most cases, a waste product [15]. The major component of the renewable biomass of relevance to SCP production is the carbohydrate fraction.

Carbohydrate substrates are divided into two major classes: saccharides and polysaccharides (Table 12). Examples of raw materials containing saccharide substrates are molasses, whey, sulfite waste liquor and wastes of fruits and vegetables. Polysaccharides substrates are of two types: starchy and cellulosic (Table 12). Starchy materials, such as grains or cassava, are usually chemically or enzymatically treated in order to convert the starch to assimilable sugars. Cellulosic materials, although abundant, normally require hydrolysis of the cellulose to saccharides. Typical examples are wood wastes, bagasse, corn cobs, oat hulls, and a significant part (40–60%) of municipal solid waste [364]. It should be noted that some microorganisms can utilize untreated polysaccharide substrates, as is discussed below.

The carbohydrate substrates most used for microbial protein production are agricultural wastes or byproducts. Advantages cited for the use of carbohydrates are their abundance, theoretical inexhaustibility, and low cost. Their use often yields a bonus by relieving disposal problems. On the other hand, these materials suffer from fluctuations in availability and cost due to seasonal harvests and climatic conditions. In addition, stores are dispersed over a large region ("biomass occurs in small piles") and the cost of collection adds considerably to raw material cost [364–366].

Table 12. Sources of biomass (renewable resources)

Polysaccharides		Saccharides
Lignocellulose	Starch	
Wood chips	Corn	Corn syrup
Crop residues	Wheat	Molasses: cane, beet
Forest and mill residues	Potato	Raw sugar juice
Urban refuse	Cassava (manioc)	Sulfite waste liquor
Animal manure	Sago palm	Whey
Wastes-vegetables, fruits	Barley	Wastes: vegetables, fruits, fermentation
	Rice	

1.2.3.1 Saccharide Substrates

Saccharide substrates are polyhydroxy aldehydes and polyhydroxy ketones with the empirical formula $(CH_2O)_n$, or their derivatives. Monosaccharides, or simple sugars, consist of a single polyhydroxy aldehyde or polyhydroxy ketone unit. The most abundant monosaccharide is the six-carbon (hexose) sugar, D-glucose, which is also the building block of the most abundant polysaccharides. Other important hexoses are fructose, galactose, and mannose.

Pentose (five-carbon) sugars occur in plants both as free sugars and in the form of polysaccharides (pentosans such as hemicellulose). Most important in nature, from a quantitative point of view, are D-xylose and L-arabinose found in the plant polysaccharides.

The most common disaccharides are sucrose, found in sugar cane and sugar beet, maltose, which is formed as an intermediate product of the action of amylase on starch, lactose, found in milk but otherwise not occurring in nature, and cellobiose, formed as an intermediate product of cellulose hydrolysis (see below). The trisaccharide raffinose occurs in reasonable quantities (1%) in beet molasses.

Molasses

Molasses is the crudest form of sucrose, being the concentrated mother liquor left after crystallization of sugar from sugar solutions. The main sources of molasses are obtained as byproducts of cane- and beet-sugar refining.

The composition of beet molasses is approximately: 48.5% w/v sucrose, 1% raffinose, 1% invert sugar, 10.8% ash, 20.7% organic non-sugars, 18% water, and 1.5 to 2.0% nitrogen. Cane molasses differs in composition from beet molasses mainly in its content of invert sugar: 33.4% w/v sucrose, 21.2% invert sugar, 9.8% ash, 19.6% organic non-sugars, and 16% water [367].

Molasses is the most abundant and readily available saccharide substrate for microbial protein production. Its use for this purpose is widespread and well developed. Both cane and beet molasses are seasonal products, varying in composition from year to year and from locality to locality. They require limited pretreatment for use in microbial protein production. Simple chemical treatments are used to reduce the mineral content and to remove excess suspended organic matter. Beet molasses must also be heated and aerated to remove SO_2 [364].

It should be noted that the rapidly growing demand for molasses for incorporation in liquid animal feeds, as well as its widespread use in other fermentation processes for manufacture of chemicals, reduces availability and increases the cost of this raw material for microbial protein manufacture.

The principal carbon and energy source for the production of baker's yeast (*Saccharomyces cerevisiae*) is molasses [368]. The fermentable sugar concentration must be between 50 and 55% (a Brix of about 80°). The pH of molasses is in the range of 6.5 to 8.5. Due to its high invertase activity, yeast ferments and assimilates sucrose rapidly in the form of invert sugar (glucose plus fructose). The fructose moiety of the trisaccharide, raffinose, is fermented by baker's yeast. The residual melibiose moiety is fermented only by some yeast strains. Other compounds in molasses, such as acetic acid, lactic acid, succinic acid, tartaric acid, and glycerol can also be assimilated. Some amino acids can serve as both carbon and nitrogen sources though in practice, most of the required nitrogen is supplied by added ammonium salts, liquid ammonia, or urea. Baker's yeast requires biotin for growth; cane molasses supplies ample amounts of biotin (0.5 to 0.8 ppm), whereas beet molasses does not (0.01 to 0.02 ppm). Therefore, at least 20% of cane molasses has to be blended with beet molasses in the preparation of the feed wort, or the feed has to be supplemented with synthetic biotin [39].

In commercial practice, yeast concentrations of 4 to 6% (dry wt.) are obtained. Although the usual biomass yield is 0.5 g cell dry wt/g substrate used, higher values of 0.52 [369] and 0.54 [370] have been reported. For fermentations with incremental feeding Wang et al. [371] found that 0.08 g of sugar was required by 1 g of yeast cell solids per h for maintenance.

Fermentation is carried out in a fed-batch manner although baker's yeast can be produced in continuous fermentation, permitting better utilization of the total available sugar and better automatic control of the fermentation process (these aspects have been adequately described by Wang et al. [371], Burrows [372], Reed and Peppler [373], Harrison [374], and Sato [375]). It should be pointed out that the technology employed in production of baker's yeast is directed to obtain a product which has good functionality in the making of bread and not as a protein source for food and feed. Moreover, inactive dried baker's yeast is too expensive for use in the animal feed industry. However, it is used as a nutritional supplement. In 1977, the estimated world's production of baker's yeast was 187,700 dry tons [13].

Large-scale cultivations of *Candida (Torulopsis) utilis* on beet molasses were carried out, in Germany during both World Wars, and on cane molasses in Jamaica (sponsored by the U.K. Government) in 1946 [8]. Both SCP products were intended for use as supplements to human food rather than animal feedstuffs, but these programs were stopped as soon as reasonable food supplies became available.

A small scale plant for SCP production is operated by Jafco Product, Inc. of Mississippi, who is growing yeast on molasses as part of liquid animal feed production. The concentrated cell suspension is incorporated, without drying, into animal rations and provides a high protein ingredient in the product [8, 376].

In Guatemala, Rolz and colleagues have investigated the growth of a number of yeast strains on cane molasses [377], and the growth of the filamentous fungus *Verticillium* sp. on cane blackstrap molasses (CBM) [378]. The fungus was grown in a 19-m^3 fermentor in a continuous culture operation (dilution rate of 0.05 h^{-1}), under aseptic conditions, on a 1% w/v CBM medium supplemented with 1% w/v of ammonium phosphate, pH 3.5 at 33 °C. The average yield (4 runs) was 0.47 g cell dry wt/g sugar utilized; 65% of the carbon present in the carbohydrates was recovered in the dry cell mass.

Commercial operations utilizing molasses are widespread with major developments in Russia, Taiwan, Cuba, and South Africa, but no details are available [366].

Spent Sulfite Liquor (SSL)
Spent sulfite liquor (SSL) is a byproduct of the paper pulping industry, and is a widely available carbohydrate substrate. It is formed by digesting wood chips in an aqueous acid solution containing magnesium, sodium, ammonium, or calcium bisulfite. During the process lignin combines with the bisulfite ion to form soluble lignosulfonate, whilst over half of the non-cellulosic carbohydrates in the wood are hydrolyzed to simple sugars. The contents of SSL were shown to vary with type of wood digested, site of collection at the paper mill, and the reagents used in the cooking process [155]. Liquors from hardwoods contain about 2 to 3% of fermentable sugars of which about 80% are pentoses and 20% hexoses. Conversely, liquors from softwoods contain about 80% of the sugars in the form of hexoses and 20% as pentoses. Spent sulfite liquor contains appreciable amounts of acetic, galacturonic, and formic acids.

Sulfite waste liquor from paper mill operations contains concentrations of SO_2 or sulfite that inhibit microbial growth. Both treating with lime, which results in the precipitation of calcium sulfite, and steam stripping in a tower are effective methods for reducing inhibitory concentrations of sulfur compounds [7]. In addition, an acid treatment may be required in order to hydrolyze xylose polymers into available xylose [155].

The amount of waste effluent produced each year by various sectors of the pulp and paper industry is enormous. Thus, sulfite pulping in the USA alone accounts for nearly 217 million tons of organic waste material annually [379]. Due to its high biological oxygen demand (BOD) (25,000 to 50,000 ppm) and chemical oxygen demand (COD), the disposal of SSL has been a significant source of water pollution in many areas. In trying to alleviate this problem, investigators have studied the possibility of using microorganisms to convert the fermentable substrates in SSL into usable products, such a propionic, acetic, lactic and formic acids, thiamin, vitamin B_{12}, acetone and butanol [379]. By far the most significant product synthesized from SSL is microbial protein. However, although 90% of the BOD can be removed from SSL by fermentation, the decrease in total organic carbon is low, due to the high lignosulfonate content. Invariably, further treatment of the fermented liquor is required before it can be discharged to water sources. Normal practice is to evaporate and burn the fermented wash or convert the residue into lignin products [366].

In Germany during World War II, *Candida utilis* (*Torula* yeast), which assimilates pentoses, hexoses, and organic acids, was produced commercially on sulfite waste liquor and wood sugar, derived from the acid hydrolysis of wood. It was used as a protein source for humans and animals. Since then, several processes using SSL as a raw material for SCP production have been developed (Tables 5 and 6). They are based on the yeast strains *Candida utilis* (SFIT-ETH, Switzerland, Rhinelander Paper Corp, USA, Boise Cascade Co., USA, Metsaluton Sellulosa Oy, Finland), and *Candida tropicalis* (SFIT-ETH, Switzerland; Mahidol University, Thailand). Although in most cases continuous fermentation is used for producing yeasts from SSL, the productivity is limited both by a low utilizable substrate concentration (2 to 3%), and the growth rate of the yeast employed. To increase productivity, an efficient, continuous, external recycle-fermentation system was developed at MIT [154]. *Candida utilis* cells were sedimented continuously with a flocculating agent, bentonite, before being recycled to the fermentor. A computer monitoring system, based on material balancing techniques (see later in the monograph) was developed to monitor and control the recycle process on-line. This system improved fermentation production significantly. A cell concentration of 65 g/liter was achieved in the recycle stream and a concentration of 25 g/liter was obtained in the fermentor. Productivity reached a maximum of 6.1 g cell dry wt/liter · h; an improvement of 150% over that reached without cell recycle.

The largest mold-SCP process now being operated is the "Pekilo" process. It utilizes SSL and was developed at the Finnish Pulp and Paper Institute [77, 380]. This process is described in detail in a later section.

Whey

Whey from cheese-making is another potentially attractive raw material for microbial protein production. Whey is obtained either from whole milk, during cheese preparation, or as the dialyzate after removal of proteins (mostly casein) by ultrafiltration (permeate). It contains about 5% lactose, 0.8% protein, 0.7% mineral matter, from 0.2 to 0.8% lactic acid, and vitamins. The amount of whey derived during cheese manufacturing from curds is huge, though it varies for different types of cheese; approximately 10 kg milk are used to produce 1 kg cheese and 9 kg whey waste [381]. In 1975, 11×10^6 tons of whey were produced in the USA and the world's production capacity was 50×10^6 tons [382]. In 1981, a production of about 26×10^6 tons of liquid whey has been estimated for the USA and the world's production capacity increased to 92.8×10^6 tons [383]. The steady increase in dairy product consumption together with the proliferation of the restrictions on direct disposal to the environment of this high BOD material, have added advantages for using whey as an SCP substrate [364]. However, commercial cheese production is limited to the most highly developed economies where many protein sources compete. Even in these countries there is insufficient quantity of whey in one place to warrant the cost of a fermentation plant, and since whey consists of about 94% water, it is too expensive to transport. An added problem is the seasonal variation in the volume of milk production.

Only a few microorganisms, such as *Kluyveromyces (Saccharomyces) fragilis* [80, 150, 384, 385], *Kluyveromyces lactis* [386], *Trichosporon cutaneum* [228], another 13 yeast strains [148], *Wingea robertsii* [148], and *Penicillium cyclopium* [81] can use lactose as a carbon and energy source for growth (see Tables 5 and 6) [387]. There is an advantage of hydrolyzing lactose, since lactose itself is a rather insoluble and not very sweet sugar, whereas its constituents (glucose and galactose) are sweeter, more soluble, have improved humectant properties, and are more readily absorbed and metabolized than the parent disaccharide. Immobilized enzyme (lactase, β-galactosidase) processes have been developed for converting lactose in cheese whey to a glucose-galactose mixture that can be utilized further by other yeast strains [388, 389].

Besides lactose, the lactic acid present in whey (0.2% in sweet whey and 0.6% in acid whey) is very rapidly used by *K. fragilis* and disappears in the early phases of growth. This organism can use 25% of the total nitrogen present in whey (which consists of 0.14% nitrogen) [390]. The yeast uses about one-half of the ammonia nitrogen fraction of the whey, and about one-half of the organic nitrogen which is composed of proteins, peptones, amino acids and other substances. Addition of a nitrogen source such as $(NH_4)_2SO_4$ phosphate and yeast extract, improves yeast growth which reaches an optimum level at pH 5.1 [84, 150, 382, 384]. Other researchers prefer to maintain the pH between 3.5 to 4.5 during fermentation in order to prevent contamination and increase foam stability [80, 228, 387, 391–393]. The heat-and acid-precipitable whey proteins, β-lactoglobulin and α-lactalbumin (57 and 21% of whey proteins, respectively) are not utilized by *K. fragilis*.

One of the advantages of yeast fermentation is the large reduction of waste that may be achieved. Amundson [391] reported a reduction of 85% of the chemical oxygen demand (COD) of whey (initially about 65,000 ppm). However, the operating conditions (lactose concentration, aeration, and temperature) associated with the maximum reduction of COD are different from those yielding a maximum amount of biomass. A study was directed to establish the relationship between these parameters [151]. The data obtained can be used in the analysis and optimization of a great number of SCP production processes from different wastes.

An elegant process is that of Amber Laboratories [80] which produces yeast SCP and/or alcohol from whey by a closed-loop system yielding a fermentation process with no polluting effluents. In this scheme, the concentrated whey is diluted with condensate water, derived from the evaporation of the fermentation broth. Since the entire fermentation broth is spray dried, no waste streams are produced.

The process operated by Amber Laboratories, with a production capacity of 5000 tons/annum, is one of the largest plants for SCP production based on whey as the raw carbon source (Tables 5 and 6, and Fig. 14) [80]. The organism of choice in this process is a strain of *Kluyveromyces fragilis*, which efficiently converts lactose into biomass under nonsterile fermentation (the yield obtained is 0.45 to 0.52 g cell dry wt/g lactose consumed). Whey (acid and/or sweet) in the concentrated form (45–50% solids) is diluted to the appropriate lactose concentration (about 3%) and added to the fermentor together with anhydrous ammonia, yeast extract (probably as a biotin source), phosphoric acid and hydrochloric acid (used to adjust the pH to 4.5). Fermentation is carried out in a 60,000 liter aerated fermentor operated in a batch, semicontinuous or continuous manner at 30 °C. The continuous operation (dilution rate of $0.125\ h^{-1}$) is started when the cell count in the broth reaches 1×10^9 cells/ml and the lactose concentration falls to 0.5 to 0.75% (w/v). If the entire contents of the fermentor are dried, the end product (Amber yeast fermentation solubles, YFS) can be a feed grade material which contains yeast, residual whey protein, a fairly high ash content and some lactic acid. If the yeast cells are separated by centrifuging and washed, the dried end product (Amber Nutrex) is food grade material with a crude protein content of 45 to 55%. Unlike *Candida* strains, *Kluyveromyces* strains form ethanol if the fermentation is carried out under almost anaerobic conditions in high whey concentration. Thus, the plant can be run to favor either SCP or ethanol production, depending upon the market prices of the two products.

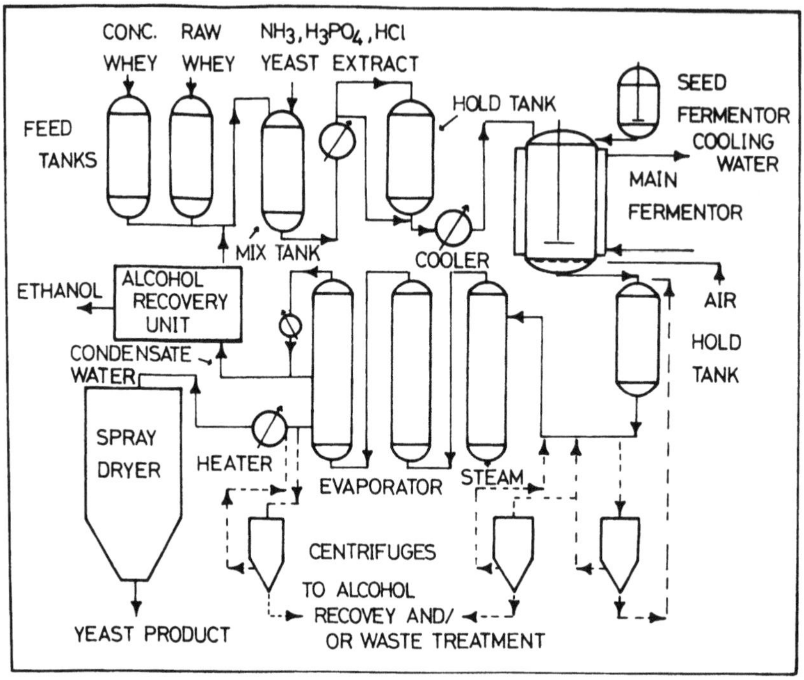

Fig. 14. Schematic for whey fermentation process. (Redrawn from Bernstein et al. [80]). (Reprinted with permission of John Wiley & Sons, Inc.)

Two-step fermentation processes for SCP production from whey have been suggested. In one of these processes the lactose is converted anaerobically by *Lactobacillus bulgaricus* to lactic acid in the first stage, followed by its utilization by aerobic thermophilic yeast (*Candida kruzei*; the "Kiel process") [394]. In another similar process, 5,6-dimethylbenzimidazole, a precursor of vitamin B_{12}, is added to the whey medium, which is then fermented anaerobically by propionibacteria, followed by an aerobic fermentation with *K. fragilis* [39]. The fermentation results in the production of propionibacteria with high levels of vitamin B_{12}, and these bacteria contribute a higher level of sulfur-containing amino acids to the biomass.

1.2.3.2 Polysaccharides

Starch

Starch is the major storage carbohydrate of plants and is produced commercially from corn, wheat, barley, rice, potatoes, tapioca, and sago (Table 12). Most starch is obtained from corn (in the USA, the annual collectable amount of corn is about 200 million tons), yielding cornsteep liquor as a byproduct. Minor sources include the tapioca starch obtained from cassava plants, and potato starch obtained from the common tuber.

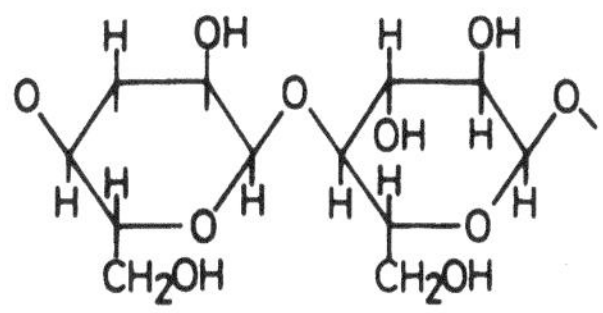

amylose

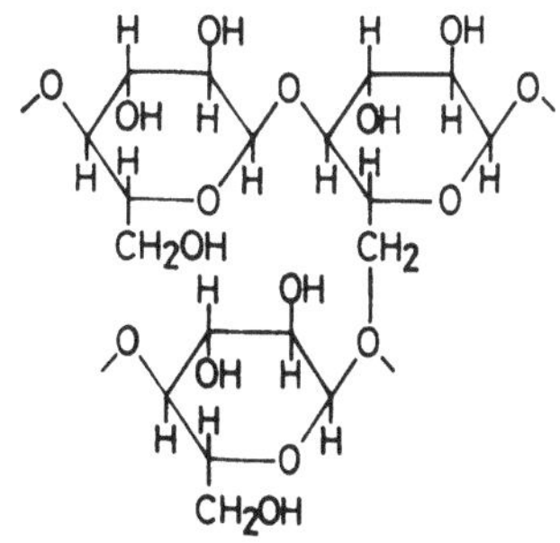

amylopectin

Starches are mixtures of two different types of polysaccharides. The minor component (about 25%) is the α-amylose molecule, which consists of long linear chains, in which the D-glucose units are bound in α(1→4) linkages. The chains vary in molecular weight from a few thousand to 500,000. Amylose is not truly soluble in water but forms hydrated micelles in which the polysaccharide chain is twisted into a helical coil with six monomer units per turn. The major component (about 75%) of starch, amylopectin, is a highly branched polymer whose molecular weight may be as high as 1 million, consisting of many short α(1→4) linked branches of approximately 20 D-glucose units each, attached through α(1→6) branch points. Amylopectin, too, yields colloidal or micellar solutions.

Starch can to a certain degree be "tailormade" to fit specific purposes and therefore have defined properties. It is possible to selectively control bulk density, absorbancy, solubility, and particle size [6]. Starch is difficult to handle owing to its insolubility in water. Its inaccessibility, especially for yeasts, is due to differing types of granule structure. Thus, starchy materials must be converted to mono- and disaccharides to be suitable as substrates for SCP production. This hydrolysis can either be catalyzed by enzymes or by dilute acids.

The major components of starch can be enzymatically hydrolyzed in two different ways. Amylose may be hydrolyzed by the enzyme α-amylase, which is more precisely called α(1→4) glucan 4-glucano-hydrolase. It hydrolyzes α(1→4) linkages throughout the α-amylose chain. Amylose may also be hydrolyzed by β-amylase, α(1→4) glucan malto-hydrolase. This enzyme breaks off successive maltose units starting from the nonreducing end of the polymer. The polysaccharides of intermediate chain length formed during β-amylase action are called dextrins. Since neither α- nor β-amylase can hydrolyze the α(1→6) linkages at the branch points of amylopectin, the end product of amylase action on this polymer is a large, highly branched "core" called a limit dextrin. The combined action of an α-amylase and an α(1→6)-glucosidase (debranching enzyme) can completely degrade amylopectin to glucose and maltose.

Several laboratory or pilot plant processes for SCP production from various starchy agricultural wastes were described (Tables 5 and 6). In most processes the starchy material is hydrolyzed chemically or enzymatically, and then fermented with a yeast strain to form the SCP product. One of these processes was based on a dual-yeast fermentation system, in which the yeast *Saccharomyces (Endomy-*

copsis) fibuliger produces α- and β-amylases for hydrolysis of starch to glucose and maltose, which enables co-growth of *Candida utilis*. Because of the faster growth of the latter yeast, the percentage of *S. fibuliger* cell solids is only 4% of the harvested biomass, which is intended for animal feed use. This "Symba" process was developed by the Swedish Sugar Corporation and operated in an aseptic manner using a 360-m^3 fermentor (400–1000 kg/day). However, the process practically was limited by the intermittent availability of the waste stream from potato processing operations, and it was finally discontinued [73, 395].

Lignocellulosic Raw Materials

The major components of agricultural residues are primarily cellulose and hemicellulose, which are the structural cell wall polysaccharides of most green plants. These are the most plentiful renewable organic compounds on earth. Since they are major constituents of agricultural as well as urban refuse, manure, and food wastes (Table 12), tremendous quantities of inexpensive cellulosic materials are potentially available for use in fermentation media. In the USA, annual collectible amounts of the major agricultural cellulosic sources are: wood chips-200, cereal straw-180, and corn stover-105 million dry tons. The potential annual amount of cellulosic material in the USA alone is estimated at 1830 million tons (including cropland, grassland and forest resources).

Pure cellulose, such as cotton fiber, is rarely found in nature. Cellulose from natural sources is frequently associated with other polymers such as hemicellulose, lignin, and pectin (a polymer containing D-galacturonic acid), which protect the cellulose fibers from the environment. Table 13 lists the composition of the lignocellulose complex for several naturally derived materials.

The molecular weight of cellulose has been estimated to vary from 50,000 to 1,500,000, equivalent to about 300 to 10,000 glucose residues. X-Ray diffraction analysis indicates that cellulose molecules are organized in bundles of parallel chains, fibrils, which are cross-linked by hydrogen bonding. Such fibrils are insoluble in water. On complete hydrolysis with strong acids, cellulose yields only D-glucose, but partial hydrolysis yields the reducing disaccharide cellobiose, in which the linkage between the D-glucose units is β(1→4). When cellulose is ex-

Table 13. Composition of cellulosic materials[a]

	Soft woods	Hard woods	Wheat straw	Bagasse	Corn stover
α-Cellulose	43.8	42.4	42.4	38.7	42.8
Hemi-cellulose	26.5	35.6	33.5	39.0	42.0
Lignin	29.5	21.7	22.5	20.6	14.0
Ash	0.2	0.3	1.6	1.7	1.2
Polysaccharides					
-C_6	56.6	51.1	45.8	46.2	49.0
-C_5	6.9	18.2	24.6	27.0	25.6
-Total	63.5	69.3	70.4	73.2	74.6

[a] Numbers are % dry, extractives-free (Courtesy of Dr. R. M. Busche, E. I. du Pont de Nemours & Company)

haustively methylated and then hydrolyzed, it yields only 2,3,6,-tri-O-methylglucose, showing that its glycosidic linkages are all 1→4 and that there are no branch points.

Hemicellulose is not related structurally to cellulose, but is a polymer of D-xylose in $\beta(1\rightarrow4)$ linkage, with side-chains of arabinose and other sugars (i.e., glucose, galactose and mannose). The amounts of hexoses (cellulose) and pentoses (hemicellulose) in different plant sources are given in Table 13.

Lignin is found in all fibrous plants and comprises from 14 to 30% of the dried plant residue. This material is the structural glue that binds filaments of cellulose into fibers for cell integrity and rigidity. Lignin is a three-dimensional polymer and its structural backbone is composed of phenylpropanoid monomeric units, linked together in a variety of complex chemical bonds (Fig. 15) [396–399]. In plants, lignin is formed by peroxidase-mediated free radical polymerization of hydroxylated and/or methoxylated cinnamyl alcohols, including coniferyl sinapyl, and p-coumaryl alcohols [396]. The relative contribution of each varies with the plant species, but in all cases lignin contains numerous and complex intermono-

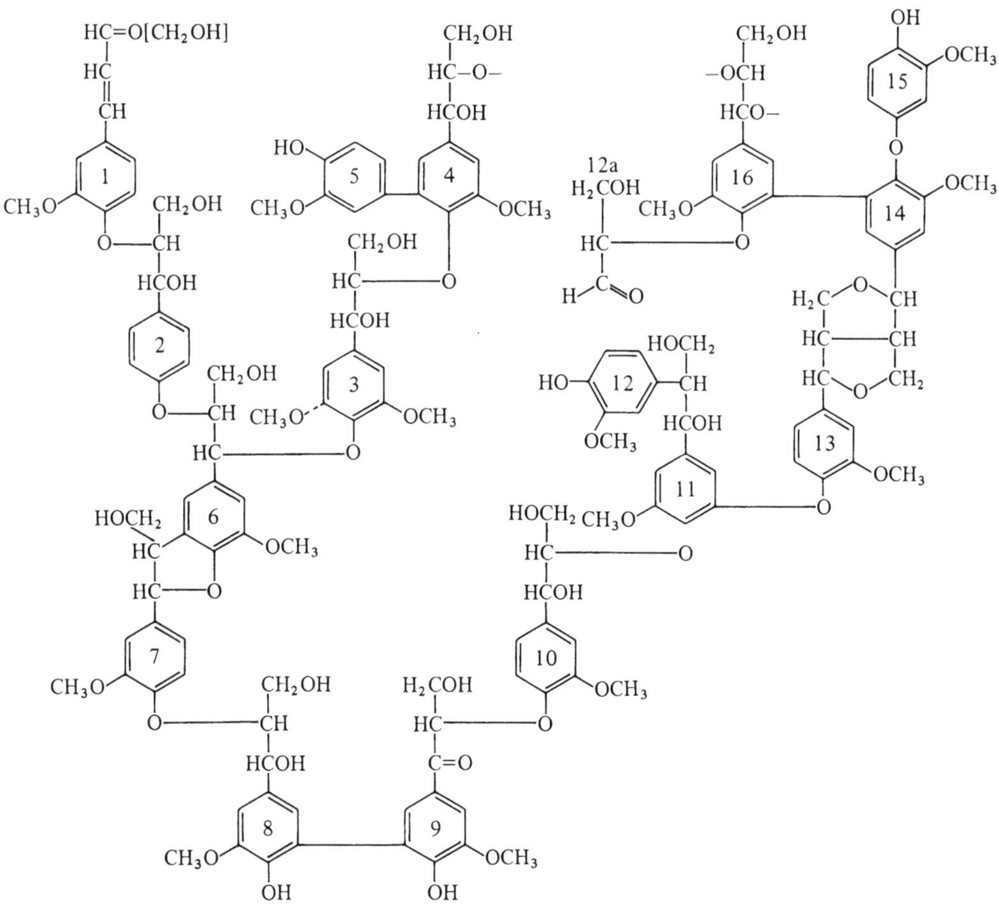

Fig. 15. Structure (approximate) of lignin

meric linkages. Although there are diverse linkages, predominant ones include the arylglycerol β-aryl ether linkage, the phenylcoumaron structure, and the biphenyl linkage [399, 400].

Based upon its aromatic structure, in theory, the controlled chemical or biological degradation of lignin can generate valuable phenolic chemicals similar in structure to oil-based phenolics now in use by the chemical industry. Several chemical processes were suggested for the production of low-molecular-weight phenolic chemicals [399]. However, this highly complex aromatic polymer is noted for its resistance to biodegradation, and only in recent years have the lignin-decomposing abilities of microorganisms been studied in relation to the potential utilization of these microbial strains for bioconversion of lignin either to useful chemicals [399], or to SCP products [401].

Most processes for the utilization of cellulosic raw materials in microbial SCP production have three stages in common. They employ some means of pretreatment in order to effect at least some initial size reduction and, more often, to cause a disassociation of lignin and cellulose. Furthermore, they involve acid and/or enzymatic hydrolysis (depolymerization) of the cellulose and hemicellulose to produce monodisaccharides. Finally, they employ fermentation to produce the SCP product. A host of technical problems are encountered in the fermentation of cellulosic materials, as depicted in Table 14.

Many different pretreatment schemes have been attempted to accomplish the required changes in native cellulosic raw materials [28, 365, 403–409]. The pretreatments are divided into physical and chemical (Table 15). The problems with these pretreatments are that they require energy, equipment, and often chemicals; they result in an irretrivable loss of sugar, and undesirable side-reactions and by-product formation. Furthermore, if acids, alkali, or organic chemicals are used, they must be recycled to minimize cost, or disposed of to prevent pollution.

Physical pretreatments can be grouped into mechanical and nonmechanical pretreatments. These methods utilize shearing and impacting forces, or other harsh external forces, to effect reduction in cellulose crystallinity and lignin content (delignification) (Table 15).

Many chemical pretreatments have been suggested as a means of lignin removal and structural modification of lignocellulosics (Table 15). The effectiveness of various physical and chemical pretreatments on the structural features,

Table 14. Technical problems related to cellulosic materials. (Based in part on Humphrey et al. [402])

1. Polymers of glucose or xylose – not readily available to microorganisms
2. Lignin and hemicellulose degradation difficult in bacteria
3. Insoluble in both a crystalline and an amorphous state
4. Difficult to agitate in suspensions above 100 g/liter cellulosic materials
5. Virtually impossible to measure microbial growth
6. Use of solid substrates difficult relative to liquid substrates in continuous cultures – difficult to provide homogenous conditions, increased sterility problems, higher power, and hence, higher capital inputs in pretreatment etc.
7. Sampling and line plugging problems

Table 15. Alternative pretreatment methods for lignocellulosic materials

Chemical methods		Physical methods	
Sodium hydroxide Ammonia Ammonium bisulfite Chemical pulping (thioglycolic acid, sodium benzoate, xylene sulfonate, nitric acid, oxygen) Sodium chlorite Gases (SO_2, NO_2, chlorine oxide)	Acids (HCl, H_2SO_4, H_3PO_4, HF) Oxidizing agent (peracetic acid, hydrogen peroxide and iron, ozone) Organic solvents (ethanol, butanol, acetone, ethylene glycol, dioxane) Cellulose solvents (Cadoxen, CMCS)	Mechanical – ball milling two-roll milling hammer milling colloid milling vibroenergy milling extrusion fitz milling grinding soaking in water	Nonmechanical – uv-radiation γ-irradiation pyrolysis steaming steam explosion freezing

crystallinity and lignin content of wheat straw was measured. It was quantified in terms of enzymatic digestibility of the wheat straw following pretreatment [410, 411]. It was found that in general:

1) chemical pretreatments were more effective than physical pretreatments,
2) multiple pretreatments were not promising, since the hydrolysis rates rarely exceeded those achieved by single pretreatment,
3) the lignin content of wheat straw is relatively more important than other structural features in controlling hydrolysis rate,
4) the subsequent hydrolysis rate increases with increasing degree of delignification up to 50% and thereafter it remains unchanged,
5) size reduction and crystallinity index reduction affect only a limited increase in the hydrolysis potential of wheat straw, and
6) NaOH pretreatment is a promising candidate for large-scale application because of its excellent capability for increasing the hydrolysis rate, for the sugar yield it renders, and for its low cost.

Also, treatments with ethylene glycol and peracetic acid are highly effective in delignifying substrate.

An interesting scheme feasible for pretreatment of lignocellulose is the acid hydrolysis process [412, 413]. Dilute acid is used to hydrolyze the hemicellulose to pentose sugar initially, and then stronger acid at higher temperatures is used to cause cellulose hydrolysis (Fig. 16).

Due to the mild hydrolysis conditions required, the hemicellulose can be removed from the cellulosic substrate before exposure to enzymatic hydrolysis. This procedure might have the advantage of increasing accessibility of the enzymes to the cellulose by the removal of the closely associated hemicellulose. Alternatively, the residues from the enzyme processing can be subjected to a two-stage hydrolysis to remove the hemicellulose before subjecting the cellulose to more stringent hydrolysis conditions. A hexose-rich prehydrolysis stream can be added to the main glucose stream for fermentation. After separation of the residual solid

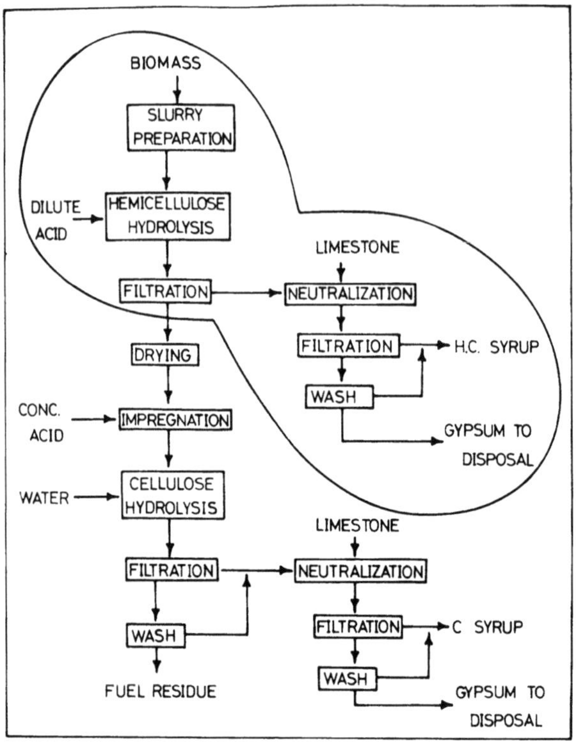

Fig. 16. Acid hydrolysis of cellulosic residues. H.C.-Hemicellulose, C-cellulose. (Courtesy of Dr. Busche)

(mostly lignin), which can be burned to provide energy, the sugar (hexose) solution is fermented to SCP. The pentose sugars also can be fermented as a separate sugar stream by the same or by other organisms.

The major disadvantage of acid hydrolysis is that the use of higher temperatures and acid concentrations, needed to increase the rate of hydrolysis of the less accessible regions of the cellulose, causes decomposition of the glucose already formed. Thus, glucose yields of about 50% of the theoretical have been realized in most cases [414]. More recently, higher glucose yields (up to 95%) have been obtained [39, 415, 416]. Additional disadvantages of highly acidic processes are that they require special equipment, and such corrosion effects a shorter working life-span.

Enzymatic conversion of cellulose by cellulases has been discussed extensively [28, 417–421]. Cellulase is a complex of enzymes containing (as in *Trichoderma reesei*), mainly endo-$\beta(1\rightarrow4)$ gluconase (Cx) (EC 3.2.1.4), exo-$\beta(1\rightarrow4)$ cellobiohydrolase (C_1) (EC 3.2.1.91) and a soluble β-glucosidase (EC 3.2.1.21). Analysis of the major enzyme groups reveals that each one consists of a number of distinct enzyme species [422–427]. At least four different endoglucanases from commercial preparations of *Trichoderma* cellulase have been described; each of the endogluconases has unique substrate specificity and biochemical properties. Different exogluconases such as cellobiohydrolase (CBH, currently being equated with C_1), cellobiohydrolase II, and exoglucohydrolase, have been isolated also, but their mechanism of action is not yet clear. The origin of the multiplicity of the enzymes with apparent duplication of function can be explained on the basis of partial hydrolysis [428], differential glycosylation of a common polypeptide chain [429], or genetic determination [430].

The current concept of how these enzymes interact to degrade crystalline cellulose suggests a primary attack by the endogluconases, followed by the release of either cellobiose (cellobiohydrolase) or glucose (exoglucohydrolase, β-glucosidase) from the nonreducing termini at the sites of attack [419, 423, 431–433]. For complete hydrolysis of insoluble cellulose, synergistic action between the components is required. Since different cellulase preparations vary widely in the proportions of the various components depending on source, growth conditions, and harvesting and handling procedures, the rate and extent of their hydrolysis of cellulose substrates vary likewise. For example, although cellulase from the fungus *Trichoderma reesei* has greater specific activity and efficiency than cellulase from other systems, its low β-glucosidase activity limits the formation of glucose from cellulose [434].

A major obstacle to fast and efficient enzymatic hydrolysis, besides the inaccessibility of cellulose, has been the inhibition of cellulase activity by cellobiose [435–438]. This may be overcome by diffusion of the formed cellobiose through membranes during hydrolysis, or addition of β-glucosidase to *T. reesei* cellulase, either in its free form or after its immobilization on hollow-fiber ultrafiltration membranes or by entrapment in calcium alginate gel beads [39, 434, 439–442]. Another possibility is shown by the simultaneous enzymatic saccharification and fermentation (SSF) process [443] employing the yeast *Candida wickerhamii* which can utilize cellobiose and thus alleviates the inhibition of cellulose hydrolysis by preventing excessive cellobiose accumulation [444–446].

In all processes that have been suggested for enzymatic hydrolysis, the cellulosic material requires some initial pretreatment (physical and/or chemical). In processes developed at the University of California, Berkeley, and the U.S. Army Natick Laboratories, the incoming cellulosic material is split into two streams; one is used to grow organisms that produce the cellulolytic enzymes, and the other is used to produce sugar for SCP production (Fig. 17) [417]. The cellulases are recovered from the first fermentation stage and added to hydrolyze both the cellulose and hemicellulose in a separate reaction. The resulting sugar solution is then passed to the fermentation stage where hexoses are converted by a yeast fermentation to SCP product. Utilization of the pentose sugars requires a separate fermentation. Residual lignin, which is removed before or after hydrolysis, is used to provide energy for the process.

Another possibility for the enzymatic hydrolysis of the cellulosic biomass is a single-step microbial conversion of cellulose. This novel process is best illustrated by the elegant approach developed at the Massachusetts Institute of Technology for ethanol production from cellulosic raw materials [447]. When cultured on ground agricultural residues such as corn stover, the anaerobic thermophilic bacterium *Clostridium thermocellum* excretes proteins with both cellulase and xylanase (hemicellulase) activities. It accumulates soluble hexoses and pentoses in

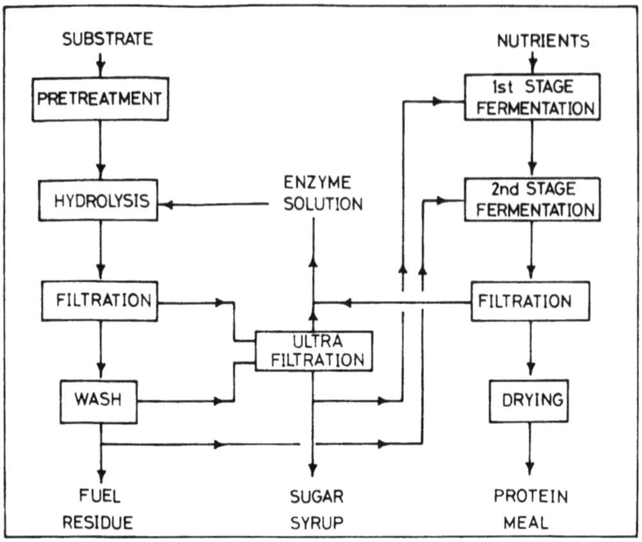

Fig. 17. Enzymatic hydrolysis of cellulosic residues. (Courtesy of Dr. Busche)

the medium while fermenting a portion of the hexoses to ethanol and organic acids. In order to consume these residual sugars and improve the ethanol yield, a second anaerobic, thermophilic bacterium, *Clostridium thermosaccharolyticum*, capable of fermenting these sugars, has been incorporated into a mixed culture. Alcohol yields of 85% of the theoretical maximum have been obtained from solka floc with mixed cultures. The advantages of this approach are a minimal requirement for pretreatment, a combined enzyme production, cellulose hydrolysis and ethanol fermentation, and simultaneous conversion of both pentose and hexose sugars to ethanol.

Similar single-step fermentation processes from lignocellulosic raw materials (after pretreatment) were developed for SCP production (Tables 5 and 6). Since direct growth of microorganisms on cellulose is slow, the loss of substrate due to the maintenance metabolism of the organism is rather high. Use of thermophiles partially overcomes this obstacle. Humphrey et al. [402] used Avicel, a crystalline form of cellulose, as the sole carbon substrate in a semicontinuous fermentation of *Thermoactinomyces* sp. at 55 °C. The preliminary results indicated that cell yields of up to 0.45 g cell/g cellulose utilized, and cell growth rates of 0.45 h^{-1} were possible. The maintenance coefficient was 0.044 g cellulose/g cell · h, consistent with that obtained for the batch glucose and cellobiose fermentations.

Combinations of two or more organisms are commonly used in cellulose fermentations (Tables 5 and 6). Pairs of *Trichoderma viride* and *Candida utilis* [448], *Cellulomonas* sp. and *Alcaligenes faecalis* (The Louisiana State University process) [449–451], *Sporotrichum pulverulentum* and *Candida utilis* [452], *Chaetomium cellulolyticum* or *Trichoderma lignorum* and *Candida lipolytica* [453], *Trichoderma viride* and *Paecilomyces fusisporus* [454], and *Cellulomonas* sp. and *Candida utilis* [455] have been used in production of SCP from cellulosic wastes. The first organism is usually cellulolytic, while the second organism is noncellulolytic. The former degrades the cellulose, and the latter grows on the degradation product, thus increasing cell mass [406].

A great amount of work has been done with lignocellulosic wastes, i.e., residues or wastes from agriculture, agro-industry, food industry, urban refuse and animal manure. Recent proposed processes are summarized in Table 6. Two main factors have caused the current developments:

1) the significant amounts of residues and wastes that are generated each year i.e., many of the developing countries with major nutritional problems produce an excess of materials high in carbohydrates (e.g., about 560 million tons of rice straw are produced annually in Asia), and
2) the rising costs for disposal of these wastes.

Until several years ago, waste (or residue) materials were assumed to have zero cost, but this assertion is no longer valid. This is due to the rising cost for collection of waste, and the realization that byproducts which have been considered wastes, are now put to various useful purposes and, therefore, have a significant cost value when intended as raw material for SCP production. Examples for such raw materials are whey and molasses (which are directly fed to farm animals or used for production of chemicals), citrus peel juice (from which pectin is recovered), and animal wastes (which can be used for biogas production). The feasibility of waste utilization, for energy generation, production of chemicals, animal feeding, or SCP production, depends on several factors, such as waste location and availability, costly waste collection and disposal situations, current usage, and value of the waste as a raw material for other products. Thus, lignocellulosic residues could be exploited as a valuable resource rather than be disposed of, destructively or otherwise, as an unwanted byproduct [400].

Although economic prospects for cellulose hydrolysis-fermentation processes are not certain, utilization of the "newly recognized resources" deserves optimistic consideration because of the reasons stated above. The resources for these processes would vary widely according to the technology and their availability in each region (Table 6).

A typical example of the use of waste material (or shall we rather say, resource) for SCP production, in a developing country, was studied by Rolz and his colleagues (Table 5) [378, 456]. The processing of coffee in Central America gives rise to byproducts, many of which are only partially utilized or wasted. A major byproduct is the water used in depulping and wasting of the coffee beans; about 15 liters of water are used for every 50 kg of coffee beans. The commercial process creates wide fluctuations in the composition of the waste water, depending on the availability of water and the amount of coffee to be processed which, in turn, varies with the season. In countries where coffee processing plants ("beneficios") are large and centralized, the waste water is recycled several times in the process due to the shortage of fresh water sources. This increases the organic matter content of the waste. Thus, when discarded into rivers, it causes large pollution problems.

A research project was initiated in the Central American Research Institute for Industry (ICAITI) in 1970, and was later evaluated in a pilot plant comprising of two 20,000-liter appropriate fermentors, located at Beneficio Curazao in San Salvador, El Salvador. When the true fungal strains, *Trichoderma harzianum* Rifai and a *Verticillium* sp. were selected, they were grown under simple, nonaseptic, batch fermentation conditions, at an initial temperature of 28 °C, which increased during fermentation to 40 °C. The pH was 3.5 initially, and it decreased to values close to 2, thus minimizing contamination by other microorganisms. An average reduction of 62% in the chemical oxygen demand of the waste was achieved; the dry mass yield was about 0.4 (g cell dry wt/g carbohydrate), with a net increase of 3 to 3.5 g of dried cell mass/liter attained after 24 h. Amino acid analysis of the product revealed a typical fungal protein profile, with limited methionine and high lysine content, that could be used to complement and balance cereals in the preparation of animal feed concentrates. Since coffee is a seasonal crop, the use of blackstrap molasses was proposed as an alternative carbon source to complete the year-round operation of the plant [378].

The low-technology SCP process thus developed can be installed at the "beneficios", so that the SCP production plant saves the water treatment costs. This was suggested by the Tate and Lyle research group in England for areas where skilled technology is not available but different raw materials are found abundantly (Table 5) [457, 458]. These processes which are characterized by low capital investment and much human involvement are particularly intended for developing countries. This concept is also illustrated by the "Waterloo" process which will be discussed later in detail.

1.3 Other Nutritional Requirements for Biomass Production

In addition to the carbon source, all microorganisms require sources of energy, nitrogen, minor elements (P, K, S, Mg), and trace elements, and may require supplemental nutrients such as vitamins. The nutritional needs of microorganisms are diverse as they differ in their inherent ability to synthesize essential growth factors from simple nutrients.

1.3.1 Energy Sources

Energy is made available to microorganisms in several different ways. Photosynthetic algae and bacteria possess photosynthetic pigments which allow utilization of solar energy (see Fig. 3). Others, called chemolithotrophs, obtain energy by oxidation of specific inorganic substrate (e.g., hydrogen, sulfur and inorganic sulfur compounds, ammonia, nitrites, and ferrous salts). These include bacteria belonging to the genera *Thiobacillus, Nitrosomonas, Nitrobacter, Hydrogenomonas, Alcaligenes* and *Desulphovibrio*. Lastly, the largest group of microorganisms – heterotrophs (or chemoorganotrophs) – gain their energy by catabolism of organic substrates.

In aerobic metabolism, oxygen acts as the final electron or hydrogen acceptor. Besides this role, oxygen may, by means of an oxygenase enzyme, be incorporated into carbon substrates in their catabolism. The oxygenase function is observed in catabolism of hydrocarbons and aromatic ring compounds. With the use of reduced carbon sources (such as hydrocarbons and methanol), oxygen is required not only to produce energy, but also to raise the oxidation levels of the substrate to that of the biomass. Therefore, the production of a unit biomass from reduced carbon sources requires much more oxygen than production from carbohydrate.

1.3.2 Sources of Nitrogen

Microorganisms show great diversity in their ability to utilize nitrogen sources. Many microorganisms are autotrophic for nitrogen, being able to grow on nitrate, ammonia, urea, and sometimes on gaseous nitrogen as the sole source of nitrogen, but others need this element to be supplied as amino acids (protein hydrolyzates), or as purine and pyrimidine bases. Microorganisms requiring amino acids may be nutritionally deficient, either being unable to synthesize amino groups or failing to synthesize certain amino acids [367].

The nitrogen is metabolized to provide mainly protein, nucleic acids, and cell wall polymers, and it constitutes up to 12% of dry weight of bacteria and 10% of fungal dry weight [459]. It is important to adjust the supply of the nitrogen source so that a ratio of 10:1 or less for carbon to nitrogen can be maintained in the medium during growth. This minimizes accumulation of lipids or storage substances, such as glycogen or poly-β-hydroxybutyric acid, and favors high protein content in the biomass product (see Chap. 3).

1.3.3 Sources of the Minor Elements

Microorganisms require a number of inorganic ions in micro and millimolar concentrations for optimal growth. The role played by these ions is twofold: enzymatic – as the catalytic center of an enzyme, as an activator or stabilizer of enzyme function, or to maintain physiological control by antagonism between activators and deactivators, and structural – the ions act to neutralize electrostatic forces present in the various cellular anionic units [460]. These two roles of any ion are important in defining the optimum concentrations for growth (see Chap. 2 of the monograph, and the comprehensive review cited in [460]).

The phosphorous source for SCP production is usually supplied either as phosphoric acid or as soluble phosphate salts. The phosphate is mostly incorporated into the nucleic acids, phospholipids, and cell wall polymers, though it occasionally may accumulate in the cells as polymetaphosphate. The phosphorus content of bacterial cells is about 1.5% (of dry wt), but this amount increases at high growth rates and at low temperatures, which reflect changes in RNA and phospholipid content ([459] see Chap. 3 of this monograph).

Potassium, magnesium and sulfur are required for various specific purposes in the cells, such as coenzymes (K), active metal in chlorophyll (Mg), components of amino acids (S), and components of coenzymes (S), membranes (Mg), and RNA (K) [460].

Trace elements such as manganese, copper, zinc, molybdemum, cobalt, and boron, (i.e., those elements are required at μM concentrations) play important roles as constituents of enzymes or coenzymes [460]. Their concentrations in growth media are shown later in the monograph.

1.3.4 Vitamins

Many microorganisms are exacting in their nutritional requirements because they are unable to synthesize complex organic compounds or "vitamins" that are, in most cases, constituents or precursors of enzymes and coenzymes. The water soluble vitamins which are important for microorganisms are: thiamine, riboflavin, pantothenic acid, pyridoxine, nicotinic acid, biotin, p-aminobenzoic acid, folic acid, and cobalamin. It is important that vitamins be supplied to the medium in minute quantities (1–50 μg/liter) whenever necessary. In undefined media, yeast extract is commonly used as a source of vitamins.

2 Fermentation Processes for Microbial SCP Production

2.1 Introduction

An understanding of the relationship between organism and environment is a prerequisite for the ability to design the process conditions inside the fermentor which give high productivity combined with a high carbon conversion and a good product composition. This understanding is also needed for reliable design of any type of fermentor, and to "tailor" that fermentor to best match the economic conditions prevailing for an overall SCP plant design [461].

The major variable cost factor in the process of SCP production is the cost of the carbon source. A high cellular yield value is therefore desired, so that efficient conversion of substrate carbon to cell carbon (mainly protein) is made maximal. The extent of this conversion is largely dependent on the fundamental properties of the organism and on its response to major process variables which determine a given environment in the fermentor. This chapter is concerned with the dependence of the cellular yield value (or carbon conversion) on both biological and process variables. These variables include dilution rate, oxygen requirement, heat evolved during fermentation, pH, temperature, CO_2 backpressure, the pattern of supply of key nutrients and their concentration in the fermentor, and the maintenance metabolism requirement of the organism. The benefits of using continuous culture versus batch culture, and mixed-cultures instead of a single-strain culture are discussed. Other factors determining the economic viability of the SCP process, such as sterility requirements, type of fermentor, recycle of media, and methods used for product recovery are also summarized.

Several large-scale SCP plants are discussed in detail in order to focus on the specific problems that arise when different carbon sources are used as raw materials, and on the answers given in these cases. Some of the current research and development efforts which may lead to significant improvement in SCP processes, i.e., type of fermentor, computer controlled fermentation, genetic manipulation techniques, and the use of thermophilic organisms, are summarized.

2.2 Physiological and Engineering Process Aspects

2.2.1 Batch and Continuous Cultures in SCP Production

So far, industrial application of continuous flow culture has been mainly in SCP production, in which the continuous process using the chemostat technique has well-defined advantages over the batch process. The relative advantages are considered below.

2.2.1.1 Kinetics of Microbial Growth in a Batch Culture

A typical batch growth curve of a microorganism grown in a chemically defined medium has three distinct phases: a lag phase, an exponential growth phase and a stationary phase. The lag phase that follows inoculation of the nutrient medium is a period of adaptation during which synthesis or repression of enzymes or structural components of the cell may occur. During this period the cell mass may change without a change in cell number. The exponential growth phase is characterized by a straight line on a semilog plot of ln X (cell concentration-dry wt/liter) versus time (t). This is a period during which the specific growth rate μ (in h^{-1}) is constant and growth is thus described by: $dX/dt = \mu X$. During the course of fermentation the chemical composition of the broth changes, since substrates and nutrients are being consumed and metabolic products are being produced. An empirical relationship between growth rate and substrate concentration was suggested by Monod [462] and has the form [1]

$$\mu = \mu_{max} \frac{S}{K_s + S}, \quad (2)$$

where μ_{max} is maximum specific growth rate (h^{-1}), S is the substrate concentration (g/liter), and K_s is a constant equal to the substrate concentration when $\mu = 0.5\ \mu_{max}$. The conversion of a substrate to cell mass may be quantified by the yield coefficient (Y_s) expressed as the mass of cells formed per unit mass of nutrient consumed; $Y_s = X/S$.

At some point, the growth rate begins to decrease, either because of the depletion of an essential nutrient or because an inhibitory product accumulates. The stationary phase occurs when all the cells have stopped dividing or when viable cells have reached equilibrium with dead cells. The total cell mass may stay constant or cell lysis may occur and the cell mass will fall.

For single cell protein production it is essential to obtain maximum biomass at the highest yield and in the shortest time. This is expressed by the term volumetric productivity (P). Productivity is the amount of biomass (g cell dry wt) produced per liter per hour. In a batch process it is necessary to calculate the productivity over the entire processing time, which includes not only the fermentation time but also the turnaround time, i. e., the time required to empty the fermentor

1 The symbols used here for equations related to batch and continuous cultures are according to Wang et al. [463]

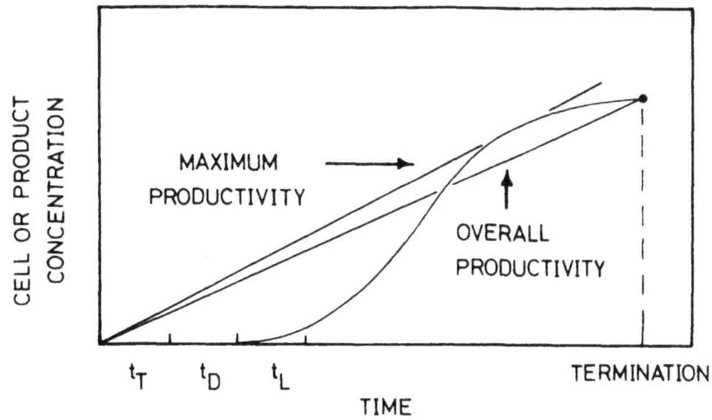

Fig. 18. Productivity in batch culture: t_T, t_D, t_L, are the culture turnaround, delay, and lag times, respectively. (Redrawn from Wang et al. [463]). (Reprinted with permission of John Wiley & Sons, Inc.)

from the previous run, wash the vessel, and sterilize new medium (Fig. 18). The overall productivity is represented by the slope of a line from the origin to the point of termination of the fermentation. The maximum productivity is given by a similar line through the origin but tangential to the growth curve; this occurs at a cell concentration less than the maximum. Thus, the extent of the improvement in productivity will depend on the length of the lag phase (e.g., using a larger and actively growing inoculum will shorten the lag phase), the turnaround time and the duration of the exponential growth phase.

2.2.1.2 Aspects in Continuous Culture Operation of Relevance to SCP Production

Continuous culture techniques based upon theoretical and experimental formulations [459, 464–470) have been widely used for approaching problems in microbial physiology [471–475]. A continuous culture is one in which a fresh nutrient medium is continuously fed into a well-stirred culture vessel, and at the same rate, the broth containing cells is continuously removed from it, such that the volume of the culture in the growth vessel is kept constant (Fig. 19).

The theory of the continuous culture implies that growth is limited by the concentration of only one substrate during the prolonged operation period (a chemostat culture), and is based on the material balance for the synthesis of cell mass in a continuous process, i.e., the equations describe the relation between concentrations of cell mass and the limiting nutrient to the flow rate of the fresh medium. The net change in concentration of cells in the culture broth in the growth vessel as a function of time can be expressed as:

$$\begin{array}{ccccccccc} \text{Increase in cell concentration} & = & \text{cells input} & + & \text{cell growth} & - & \text{cells output} & - & \text{cell death} \\ \dfrac{dX}{dt} & = & \dfrac{F_1}{V}X_0 & + & \mu X & - & \dfrac{F_2}{V}X & - & \alpha X\ . \end{array} \quad (3)$$

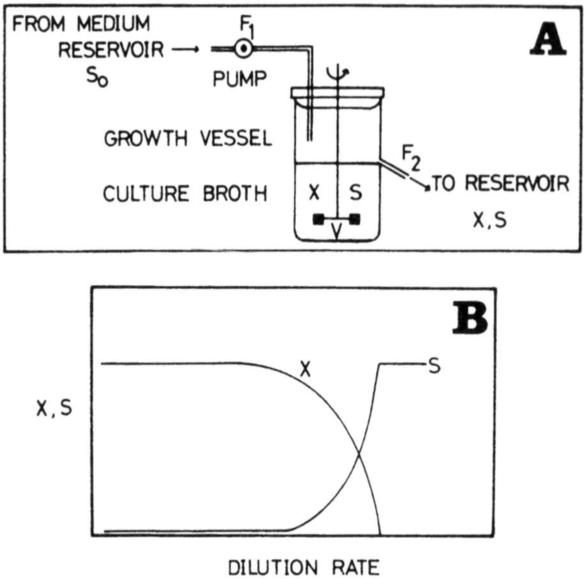

Fig. 19. A. A schematic diagram of a chemostat. **B** Theoretical continuous culture isotherms – the relation of cell concentration (X) and limiting nutrient concentration (S) dilution rate (D). (Redrawn from Goldberg and Er-el [476]). (Reprinted with permission of Process Biochemistry)

Where F_1 is the medium inlet flow rate (liter/h), F_2 is the culture broth outlet flow (liter/h), V is the volume of the culture broth in the growth vessel (liter), X_o is the cell mass (g/liter) in the feed, and α is the death rate (h^{-1}). Since $F_1 = F_2$, $X_o = 0$, $\mu > \alpha$, and if D is the dilution rate ($F_2/V;h^{-1}$), it follows that:

$$\frac{dX}{dt} = \mu X - DX = X(\mu - D)\,. \tag{4}$$

At a steady state of growth, $dX/dt = 0$, and therefore the specific growth rate is equal to the dilution rate ($\mu = D$); the concentration of the cells remains constant with time. The growth rate of cells is then limited by the supply rate of the growth-limiting nutrient to the culture. The relationship between the specific growth rate and the concentration of the limiting substrate was suggested by Monod [462] and is given by Eq. (2).

The net change in substrate concentration in the culture broth is given in the following material balance Eq.:

$$\begin{array}{ccccccccccc} \text{Substrate accumulation} & = & \text{Substrate input} & - & \text{Substrate output} & - & \text{Substrate consumption} & - & \text{Maintenance requirement} & - & \text{Product formation} \\ \frac{dS}{dt} & = & DS_0 & - & DS & - & DX\frac{dS}{dX} & - & m_sX & - & \frac{q_pX}{Y_{P/S}}\,. \end{array} \tag{5}$$

Where S_o and S are the inlet and outlet concentrations of the limiting nutrient, m_s is the maintenance requirement, q_p is the specific rate of product formation (g product/g cell dry wt/h) and $Y_{P/S}$ is the conversion yield of substrate to product. Frequently, the maintenance requirement is low relative to growth ($m_sX << DXdS/dX$, when μ approaches μ_{max}), and in the case of SCP production, product formation is negligible as compared to that part of substrate which is incorporated into the biomass. Adding the yield factor Y_s into the last equation, the steady-state ($dS/dt = 0$) for the biomass concentration can then be expressed as:

$$X = Y_s(S_o - S) \tag{6}$$

and for the concentration of the limiting substrate in the culture broth we can write:

$$S = K_s \frac{D}{\mu_{max} - D}. \tag{7}$$

$$X = Y_s\left(S_0 - \frac{DK_s}{\mu_{max} - D}\right). \tag{8}$$

Using Eqs. (7) and (8), the theoretical behavior of a chemostat may be described, as is shown graphically in Fig. 19 B.

From Eqs. (6) and (7), it can be seen that in steady-state conditions the concentration of the limiting-substrate is independent of its concentration in the medium reservoir, but dependent on the dilution rate. At the same time, the cell concentration becomes a function of the substrate concentration in the medium reservoir. Any fluctuations which disturb the steady-state, e. g., an increase in the concentration of the limiting nutrient upon its addition to the culture broth, will set up opposing reactions (an increase in cell concentration) that will tend to restore the steady-state. It is this inherent self-balancing property of the chemostat culture that makes this form of continuous culture easy to operate over a long period of time [469].

2.2.1.3 Advantages of Continuous Culture over Batch Fermentation

A continuous fermentation is recommended for development and production of high-quality biomass for the following reasons:

1) Highly defined growth conditions – constant conditions, other than the parameter which is being changed, allow easy observation of the effects of the independent variable, such as use of chemostat (supply of one nutrient in limited amounts, whilst all others are present in slight excess), dilution rate and/or specific growth rate may be manipulated as can parameters such as temperature, pH and ionic strength.
2) Efficient utilization of the carbon source – chemostat experiments performed yield results which can be used to calculate how to supply nutrients at optimal

feed rate – effects both higher yields and better use of the carbon source than in batch mode [265, 477]; reduces cost and improves process viability.
3) Continuous culture productivity greater than that of batch culture – (expressed in g of cell dry wt per unit volume in the fermentor per unit time) [463, 478] saves on fermentor volume and processing capacity, requires less capital investment. In addition, reduction in turnaround time obviously reduces operation costs.
4) Much longer growth period than in batch culture – due to steady-state maintenance.
5) Greater control and monitoring ability than in batch culture – enhances the use of an on-line computer.
6) System and product more homogenous than those observed in batch mode.

The above discussion dealt with a single-stage chemostat, the basic design of most continuous culture systems. New types and concepts of continuous culture systems have been developed, and the reader is referred to the reviews of Herbert [479], Harrison [480], and Wang et al. [60]. Here, only two modifications will be described briefly: 1) cell recycle and 2) study of dynamic responses to transients. Recycling of cells from the chemostat effluent provides a means of continually inoculating the fermentor; this adds stability to the system and minimizes the effect of process perturbations. At the same time it allows a chemostat to be operated at higher steady-state cell concentrations than in a chemostat without recycle. As a consequence of cell recycle, it is possible to increase overall productivity of the system because a dilution rate greater than the maximum specific growth rate may be employed. Also, increased cell concentration in the culture vessel permits a proportionately increased rate of substrate consumption [463].

An interesting development in the concept of continuous culture has been the treatment of transients between steady-states to obtain growth parameters [481–486]. An understanding of these transients the dynamic responses of the continuous culture to perturbations, is necessary for the design of effective process control systems. The ability to induce transients in continuous culture, i.e., by changes in pH, flow rate, temperature or concentration of the limiting-nutrient, can be a useful tool in investigating the physiology of growth and product formation.

2.2.1.4 The Applications of Continuous Culture in Research Relevant to SCP Production

Applications of continuous culture in the selection and enrichment, as well as in physiological and biochemical studies of microorganisms that are of relevance to SCP production are well documented (see the review of Goldberg and Rokem [487]).

Selection and Enrichment

The following brief account illustrates the various approaches used in experimentation with continuous culture in this field.

– *Dilution rate* – When the continuous culture is operated with a fixed dilution rate, there is a selection of microorganisms which can grow with maximum specific growth rates (μ_{max}) equal to the dilution rate employed. For example, a continuous process with increasing dilution rates served as a selection method for the enrichment of fast-growing, methane-oxidizing bacteria, and resulted in the isolation of a single obligate methylotrophic bacterium [488]. Fast growing strains of *Saccharomyces cerevisiae* were selected by employing very rapid flow rates in a continuous culture with sucrose as the carbon source [489].
– *pH* – By gradually decreasing the medium pH during the course of continuous cultivation for prolonged periods, one can select bacteria capable of growth in a medium of greater acidity [488, 490].
– *Temperature* – The use of relatively high temperatures in continuous cultures, allows the selection of thermophilic bacteria [491] and thermotolerant yeast strains [492] capable of utilizing substrates with a low level of oxidation, which require correspondly high heat inputs in order to be oxidized, e. g., methanol.
– *Isolation of strains and mutants* – Various bacterial strains were selected in the continuous culture for growth on a particular carbon source such as formaldehyde [347], methane [493], H_2+CO_2 [494], and methanol [496]. Isolation of mutants is also possible in the continuous culture because the mutant obtained is able to utilize the growth-limiting nutrient more effectively, thus outgrowing the parent.

Using three sequentially connected continuous culture vessels and adding a mutagen to the first stage, a methanol-utilizing yeast was isolated from a culture collection of various yeast strains [496].

These few examples demonstrate the potential of the continuous culture for the selection of, usually, the microorganism that can best compete with others, and is best able to utilize the limiting growth substrate. The continuous culture thus provides a constant selective pressure of parameters that are most important to effect an efficient SCP process.

Physiological and Biochemical Studies

Using the continuous culture (chemostat) technique it is possible to grow cells with defined properties, under defined and reproducible conditions that are essential for physiological and biochemical studies. The following examples show the relevance of such studies to SCP production processes.

1) Growth on toxic carbon sources – Methanol, formaldehyde and formic acid, which are intermediary compounds in the oxidation pathway of methane to CO_2, are toxic to different microorganisms. These one-carbon compounds are present in natural gas or in methanol, which are raw materials for SCP production [497]. Growth of methylotrophic bacteria was demonstrated in the chemostat on these carbon sources as the limiting nutrients [346, 498]. This was possible because the steady-state concentration of these compounds in the culture broth was sufficiently low to prevent toxicity effects.
2) Simultaneous utilization of two carbon sources – Some of the methylotrophic bacteria are obligate (or specialist) organisms, i.e., they are unable to grow on anything besides methane, methanol, and/or methylated amines as carbon and

energy source [270, 499]. By the use of a chemostat culture it was shown that such an organism, *Pseudomonas* C, was unable to utilize formic acid as a sole carbon source [342]. However, in the presence of methanol, formic acid was completely oxidized to CO_2, i.e., it served only as an energy source, resulting in an increase of 12% in the cellular yield, calculated on the basis of the methanol utilized. In a similar way, it was shown that in the presence of methanol, *Pseudomonas* C utilized glucose only as an energy source, and not as a source for carbon, resulting in an increase of 12.5% in the yield, calculated on the basis of the methanol utilized [343]. Addition of glucose to a culture of *Pseudomonas* C growing in a chemostat on a medium containing methanol plus formic acid caused an increase in the cellular yield of 18.5%, which was higher than the maximal increase in yields obtained when either glucose or formic acid was added to a methanol-grown culture (Samuelov and Goldberg, unpublished results). This finding has an important application in the industrial process for SCP production, because the utilization of different wastes containing formic acid (industrial wastes) or glucose (molasses) in addition to methanol would increase process efficiency.

Methylotrophic yeasts, unlike *Pseudomonas* C, are able to grow on glucose as a sole carbon source [500]. When methanol/glucose mixtures [502] were used, it was shown that the methanol could be utilized by *Hansenula polymorpha* and *Kloeckera* sp. 2201 at dilution rates even higher than the maximal dilution rate possible for growth on methanol alone. This is possible due to the fact that, in the case of *Hansenula polymorpha* [503], the dissimilatory enzymes of the methanol metabolism are synthesized even at high dilution rates. This may be significant in obtaining greater productivity of yeast cells growing on methanol alone. Also it may enable the production of yeast SCP on molasses/methanol mixtures. Similar to obligate methylotrophic bacteria, yeasts use formic acid (and formaldehyde) as additional energy source, in the presence of methanol; e.g., formic acid caused an increase of 20% in cell yield of a methanol-limited *Pichia pastoris* [504].

Other applications of continuous cultures are discussed elsewhere in this monograph. These include mixed cultures, effect of O_2 and CO_2, changes in the composition of SCP product by varying the dilution rate, studying yield coefficients and maintenance requirements, variation in substrate concentration, and identification of growth-limiting nutrients other than the carbon source through medium optimization.

2.2.2 Improvement of Media for Biomass Production

In order to achieve optimal fermentation conditions for biomass production, the cells should be grown in a continuous culture with the carbon source as the limiting nutrient. Only under chemostat conditions, with a suitable medium, would a linear relationship be obtained between the concentration of the carbon substrate in the inflowing medium and the concentration of cells in the culture broth, thus ensuring maximal utilization of the carbon source. In addition, the use of exact concentrations of phosphorus and nitrogen for a given amount of the utilized car-

Table 16. Relative concentrations of elements in bacterial cells, in commonly used media and in an optimized medium

Element	Ratio (element/N) × 100		
	Bacteria [a]	Media [b]	An optimized medium [c]
N	100	100	100
Mg	4.9	15.0	7.15
K	14.0	201.0	5.65
Ca	3.0	11.0	1.1
P	23.0	176.0	32.7
Fe	0.3	2.2	0.54
Zn	0.14	0.13	0.003
Cu	0.03	0.04	0.004
Mn	0.05	0.05	0.001
Mo	0.002	0.09	0.002
B	0.006	0.01	0.0005

[a] Data are given as mean values calculated from chemical determinations of Gram positive and Gram negative bacteria [505]

[b] Data are given as mean values calculated from media used [505].

[c] Results are given for optimized mineral medium used for growth of *Pseudomonas* C on methanol [477]. (Taken from Goldberg and Er-el [476], with permission of Process Biochemistry)

bon source saves on the cost of both raw material, and of removing excesses of these compounds before discharging the waste medium.

There are two basic approaches for development of an optimal growth medium [476]. The first approach is based on the known amount of each element required for the synthesis of biomass. Thus, a method was proposed by Dostalek et al. [505] and Haggstrom [506] for designing a starting medium for bacterial growth on methanol, based on the elemental composition of a number of bacteria (Table 16). However, the concentration of minerals in growth media used by various investigators shows surprisingly wide variation, suggesting that criteria other than cell composition have been used to formulate the relative concentrations of minerals [505]. In comparing the relative compositions of bacterial cells and the corresponding compositions of media (Table 16), it can be concluded that a general model for designing a starting medium for bacterial growth apparently does not exist, especially for fastidious organisms which are able to grow only on complex media. In an optimized medium (Table 16), which was developed by the pulse method (see below), all the elements were found to be present in much lower amounts than in the different known media. Thus, the optimized medium allowed a saving on the cost of medium, in addition to minimizing the concentrations of unutilized chemicals in the waste stream.

The second approach, the pulse method, consists of identifying growth-limiting nutrients by injection into the growth vessel of suspected growth-limiting nutrients and observing the reaction of the continuous system [459, 477, 478, 507–511). By comparing a number of methods for medium optimization, which utilize a continuous culture operating under nutrient-limited conditions (chemostat), it

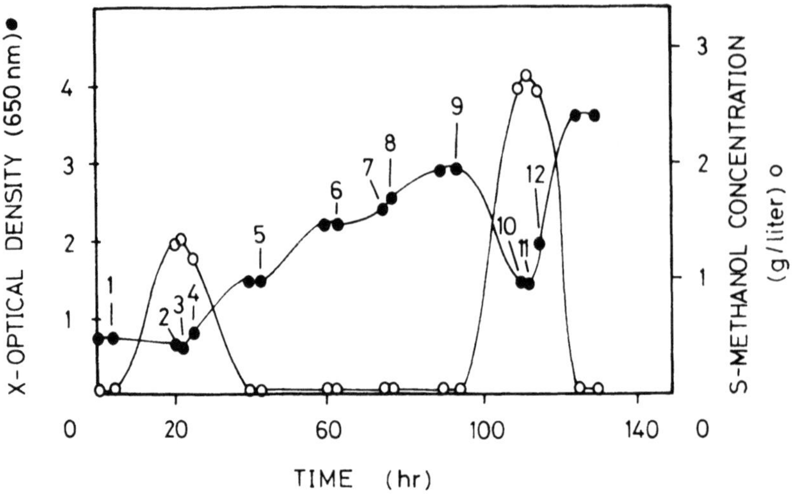

Fig. 20. Optimization of a growth medium by the continuous culture technique. *Pseudomonas* C cells were grown in a chemostat on a medium containing methanol (1 g/liter), $(NH_4)_2SO_4$ (1 g/liter), and $CuSO_4 \cdot 5H_2O$ (10 µg/liter) at 35 °C. The dilution rate was 0.35 h^{-1}, with a stirring rate of 600 rpm and an aeration rate of 0.5 volume/volume per min. Numbers indicate compounds added (g/liter) to the growth vessel (V) or to the medium reservoir (R) as follows: 1. Methanol, 1 (R); 2. $MnSO_4 \cdot 5H_2O$, 1×10^{-5} (V); 3. $CuSO_4 \cdot 5H_2O$, 1×10^{-5} (V); 4. $CuSO_4 \cdot 5H_2O$, 7×10^{-5} (R); 5. Methanol, 1 (R); 6. Methanol, 1 (R); 7. $(NH_4)_2SO_4$, 0.5 (V); 8. $(NH_4)_2SO_4$, 1.5 (R); 9. Methanol, 1 (R); 10. $FeSO_4 \cdot 7H_2O$, 0.001 (V); 11. $MgSO_4 \cdot 7H_2O$, 0.1 (V); 12. $MgSO_4 \cdot 7H_2O$, 0.1 (R). (Redrawn from Mateles and Battat, with permission [477])

is evident that they are all essentially modifications of the same basic pulse technique (see the detailed review by Goldberg and Er-el [476] on the use of the chemostat technique for medium optimization).

The example of work aimed at the improvement of a medium supporting growth of *Pseudomonas* C on methanol is shown in Fig. 20. This experiment was carried out in order to ensure that methanol was the growth-limiting substrate, and that only a low concentration of inorganic elements remained unutilized in the medium. It was found that in the original medium Cu, N, and Mg were limiting-elements, Mn and Fe were non-limiting elements, and methanol was a toxic compound. Based upon these results, an improved mineral medium was developed which supported a linear relationship between the concentration of methanol in the inflowing medium and cell concentration. At a certain stage the oxygen transfer capacity of the apparatus from gas (the air stream) to liquid (the medium), became a limiting factor (Fig. 21). A biomass yield of 0.54 g cell dry wt/g methanol utilized was obtained with the improved medium [477], as compared with a yield value of 0.31 obtained with the original medium [512].

A further result of identifying a succession of limiting nutrients is that the stoichiometric requirement for the elements can be readily calculated from the cell concentration and from the concentration of limiting nutrients. Table 17 presents such data for different methylotrophic bacteria grown on methanol. These data show that similar results were obtained when the chemostat pulse technique was

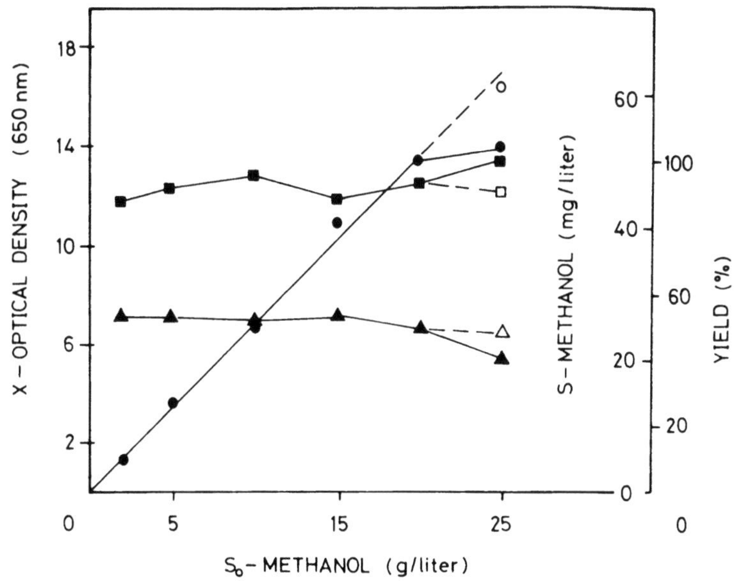

Fig. 21. Effect of different concentrations of methanol (So) on the steady-state concentrations of *Pseudomonas* C cells (X) in a chemostat culture. Cells were grown with a stirring rate of 1000 rpm, an aeration rate of 4 volume/volume per min and a dilution rate of 0.35 h^{-1} at 35 °C. ●, Cell concentration in a steady-state (X); ■, methanol concentration in the culture broth in the steady-state (S); ▲, yield value (g cell dry wt/g methanol utilized × 100). Broken lines, the aeration rate was increased to 5 volume/volume per min. (Redrawn from Mateles and Battat, with permission [477])

Table 17. Element requirements for growth of methylotrophic bacteria on methanol

mg Element required/g cell dry wt

Element	Chemical determination *M. methanolica*	Batch-culture *Methylomonas* sp. 1	Chemostat-cultures *Pseudomonas C*	*Methylomonas* sp. 2	*Pseudomonas AM-1*	Exponential fed-batch *Protaminobacter ruber*
N	139.5	95.4	110.0	–	152.0	13.9
Mg	2.8	5.0	7.8	6.4	2.1	2.8
K	3.5	957.8	6.2	21.6	–	16.4
Ca	1.9	–	1.2	0.5	–	0.4
P	26.0	439.4	36.0	41.6	16.9	51.1
Fe	1.1	0.052	0.6	0.9	–	0.14
Zn	0.1	–	0.004	0.2	–	–
Cu	0.3	–	0.004	0.008	–	–

References: *M. methanolica* [505]; *Methylomonas* sp. 1 [513]; *Pseudomonas C* [477]; *Methylomonas* sp. 2 [495]; *Pseudomonas AM-1* [514]; *Protaminobacter ruber* [508]. (Taken from Goldberg and Er-el [476], with permission of Process Biochemistry) applied for different methylotrophic bacteria, for the amount of the bulk elements (N, P, Mg) required for the production of biomass. However, it should be noted that the concentration of trace elements was different for various microorganisms, and varied greatly from the data obtained by chemical determination of the elemental composition of biomass (see also Table 16). This implies that criteria other than the chemical determination of cell composition are required when optimization of media is considered, and that the process of optimization of a growth medium should be carried out for each microorganism. The data in Table 17 also show that the use of a batch culture for improvement of a growth medium resulted in relatively high concentrations of K, and P, which were required for maintaining the pH throughout the fermentation process.

The chemostat technique for media optimization is useful not only for growing organisms on relatively simple defined media, but also for highly fastidious microorganisms, where a large number of possible nutrients are required for growth, and must therefore be tested. This is illustrated by results shown in Table 18 for growth of different *Bacillus* strains which display amino acid and vitamin auxotrophies. The optimized media developed contained glutamic acid, methionine, corn steep liquor, yeast extract and soy peptone, with complex media varying for each strain. Thus, for example, the amounts of Mn and Cu in the improved medium were higher for *B. cereus* than for *B. cladotenax* by 112- and 10,000-fold, respectively (Table 18).

It can be concluded that the chemostat technique enables saving of time and effort in the process of media optimization, as compared to batch culture techniques, and results in better knowledge of the stoichiometric requirements for the nutrients. Optimization of media was carried out both for prokaryotes and eukaryotes and, in all cases, a considerable increase in cell yield and concentration was obtained in the improved medium as compared with the initial medium [476].

Table 18. Comparison of optimized media for growth of different *Bacillus* strains

mg Element required/g cell dry wt			
Element	*B. thuringiensis*	*B. cereus*	*B. caldotenax*
N	103.8	108.6	118.9
Mg	14.4	4.2	5.7
K	7.5	101.2	35.6
Ca	0.39	0.16	0.38
P	13.2	48.0	19.0
Fe	0.015	0.093	0.046
Zn	–	0.02	0.1
Cu	–	0.01	0.000001
Mn	–	1.46	0.13

References: *B. thuringiensis* [515] – corn steep liquor, yeast extract and soy peptone were included in the medium; *B. cereus* [511] – glutamic acid was included in the medium; *B. caldotenax* [510] – methionine was included in the medium. (Taken from Goldberg and Er-el [476] with permission of Process Biochemistry)

2.2.3 Growth Yields of Microorganisms

The SCP product must have a competitive price on the protein market. Since the carbon substrate is costly, the conversion efficiency of substrate carbon to cell carbon will determine competitivity. Conversion efficiency is important due mainly to the following reasons.

1) A high conversion efficiency will result in high cellular yield and productivity. The protein yield is dependent on both the yield coefficient and the protein content of the bacterium. In any commercial process aimed at producing an SCP product, it is desirable to increase productivity by increasing the cell concentration up to the highest possible value where productivity will be limited only by the dilution rate. However, it is unlikely that either of these parameters will limit productivity but rather it will be limited by the rate of oxygen transfer to the cells.

2) The higher the conversion efficiency, the less oxygen is required per unit of cell produced (Y_0, g dry wt of cells/g oxygen consumed), and a proportionally smaller

Table 19. Growth parameters of microorganisms grown on different substrates under carbon limitation conditions

Growth substrate	Organism	Y_s	Y_s^{max}	m_s	Ref.
Ethanol	*Candida utilis* A-49	0.75	0.76	0.02	[352]
Ethanol	*Acinetobacter calcoaceticus*	0.75	0.846	0.115	[353]
Ethanol	*Saccharomyces cerevisiae*	0.63	–	0.004	[516]
Acetic acid	*Acinetobacter calcoaceticus*	0.42	0.514	0.447	[353]
Acetic acid	*Candida utilis*	0.4	0.51	0.17	[517, 518]
Glucose	*Escherichia coli*	0.50	0.53	0.09	[519]
Glucose	*Saccharomyces cerevisiae*	0.40	0.51	0.018	[519, 520]
Cellulose	*Thermoactinomyces* sp.	0.45	–	0.04	[402]
Cellobiose	*Thermoactinomyces* sp.	0.27	–	0.027	[402]
Methane	*Pseudomonas* sp.	1.00	–	0.03	[521]
Methane	*Pseudomonas* sp. (a mixed culture)	0.69	–	0.021	[521]
Methanol	*Hansenula polymorpha* CBM 11	0.41	–	0.054	[522]
Methanol	*Pichia pastoris* CBM 10	0.41	–	0.009	[523]
Methanol	*Hansenula polymorpha*	0.36	–	0.016	[524, 525]
Methanol	*Pseudomonas* C	0.54	0.61	0.08	[346]
Methanol	*Methylophilus methylotrophus*	0.53	0.59	0.12	[526]
Methanol	Bacterium 7-3	0.47	0.52	0.06	[527]
Methanol	*Pseudomonas* 1	0.38	0.46	0.048	[526]
Methanol	*Pseudomonas* 135	0.38	0.39	0.083	[526]
Formaldehyde	*Pseudomonas* 1	0.32	0.40	0.045	[346, 526]
Formaldehyde	*Pseudomonas* 135	0.27	0.35	0.081	[346]
Formic acid	*Pseudomonas* 1	0.14	0.15	0.046	[346]
Formic acid	*Pseudomonas* 135	0.15	0.17	0.20	[346]

Y_s – Yield coefficient (g cell dry wt/g substrate utilized)
Y_s^{max} – "true yield", $m_s = 0$ (g cell dry wt/g substrate utilized)
m_s – maintenance coefficient (g substrate/g cell dry wt h)

amount of heat is evolved (Y_{kcal}; g dry wt of cells/kcal evolved). This in turn effects a decrease in the specific consumption of cooling water (see below).
3) A high conversion efficiency reduces the costs of biomass separation and quantity of waste water produced.

Table 19 gives examples of cellular yields which may be obtained under good growth conditions, using continuous culture with carbon substrate-limited growth, aimed at producing cells with a high protein content. When cultures are grown in the presence of excess carbon substrate, large quantities of storage material, in the form of carbohydrates and fats, may be deposited inside the cells. Synthesis of such polymerized forms generally requires a smaller expenditure of energy than does protein synthesis and so, yields higher than those recorded in Table 19 (based on the dry weight of cells) are obtained, even though such cells are consequently lower in protein [36]. This problem will be dealt with again later in the monograph.

2.2.3.1 Theoretical Biomass Yields for Different Carbon Substrate

As a result of the great influence of yield on the economics of SCP production, it is valuable to consider the theoretical maximum yield coefficients that might be expected from different carbon substrates, in attempting to maximize the yield as a function of substrate or microorganism used for SCP production. This complicated problem was tackled in a fundamental way by Stouthamer and his colleagues, and the reader is referred to their articles [528–530].

The predictions of yields of methylotrophs growing on various C_1-substrates are of particular interest, because of subtle differences in oxidation level of series of substrates (methane, methanol, formaldehyde and formic acid), and a complex variety of known assimilatory and dissimilatory pathways (see Anthony [18] for a detailed review on this subject).

The prediction of yield requires knowledge of the ATP yield (and the degree of uncoupling) of the dissimilation and the energy requirement of the assimilation pathways, the elemental composition of the biomass, and the maintenance requirement of the organism (see below). However, the calculations made [18, 531, 532] did not take into consideration several problems. These include: 1) the excretion of partly oxidized intermediates into the medium by bacteria grown in continuous cultures under conditions of methanol limitation [477, 533, 534], 2) the operation of two oxidation and assimilation pathways in one organism, as previously discussed (Chap. 1), and 3) changes in the cellular composition due to changes in growth conditions (see later in Chap. 3).

Nevertheless, two main generalizations which were predicted from theoretical data have been fully confirmed and are thus of great relevance to SCP production. One prediction [18, 531, 532], that yields of bacteria utilizing the ribulose monophosphate (RMP) pathway should be considerably higher than those utilizing the alternative serine pathway, has been amply confirmed by experimental data [346, 526]. This is shown in Table 19 by the difference in the yields obtained for RMP-pathway methanol-utilizing bacteria such as *Pseudomonas* C and *Methylophilus methylotrophus*, as compared with serine-pathway bacteria such as *Pseudomonas*

1 and *Pseudomonas* 135. The second prediction was that relatively low yields with respect to methanol and oxygen can be expected for yeasts as compared with most bacteria. This was reasoned by the fact that the first step in methanol oxidation in yeasts involves a flavoprotein oxidase which is not coupled to ATP synthesis [18]. This prediction was also confirmed [524, 525] as is evident by comparison of the yields obtained for methanol-grown yeast, *Hansenula polymorpha*, and different methanol-grown bacteria [Table 19]. In light of the conformity between theory and experiment, it is improbable that a yeast or a serine-pathway methylotrophic bacterium would be chosen for SCP production [36]. Indeed, the largest industrial SCP production process uses a RMP-pathway bacterium (Table 4).

When considering carbon substrates other than C_1-compounds, the yield of biomass can be expressed in terms of cell carbon per gram of substrate carbon rather than a weight yield (g cells/g substrate) or molar yield (g cells/mole substrate), [535]. This form is often useful in comparing yields on substrates of different degrees of oxidation. Assuming that carbon conversion is the primary controlling factor in an aerobic fermentation where the only products are biomass, carbon dioxide and water, the total energy available for the formation of unit quantity of biomass is thus proportional to the heat of combustion of a quantity of substrate containing 1 g of carbon (heat of combustion, kcal/g of substrate carbon). Attempts have been made to correlate the yields of microorganism with the

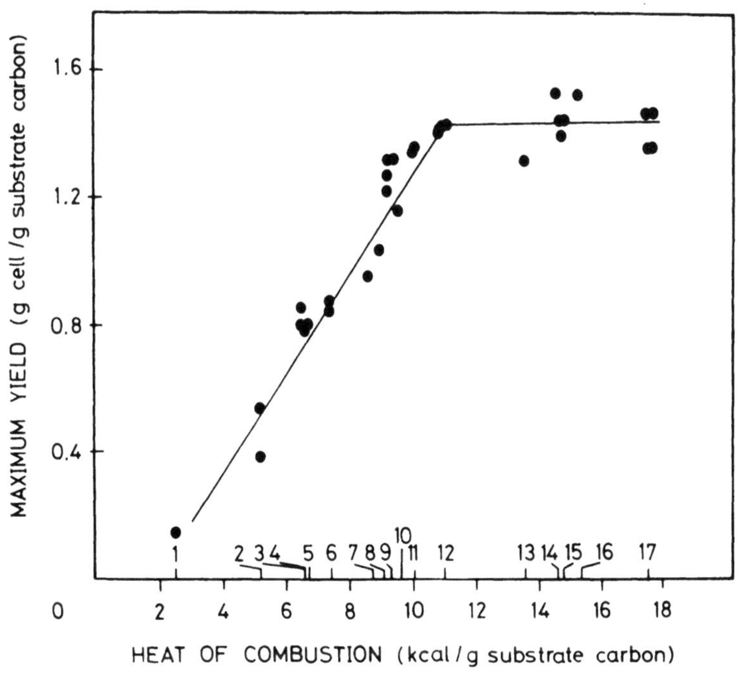

Fig. 22. The relation between growth yield and heats of combustion of various growth substrates. 1, oxalate; 2, formate; 3, citrate; 4, malate; 5, fumarate; 6, succinate; 7, acetate; 8, benzoate; 9, glucose; 10, phenylacetic acid; 11, mannitol; 12, glycerol; 13, ethanol; 14, propane; 15, methanol; 16, ethane; 17, methane. (Redrawn from Linton and Stephenson, with permission [538])

energy available through oxidation of the carbon substrate (heat of combustion) [536, 537]. Linton and Stephenson [538] found good correlation for substrates with heats of combustion up to 11 kcal/g substrate carbon, as is shown in Fig. 22. Above this apparent critical value, the yield did not increase with heat of combustion. Hence, the authors suggested that when the heat of combustion of a given organic compound is less than 11 kcal/g substrate carbon, energy generated during the biological oxidation of the compound is insufficient to reduce all the available carbon in the molecule to the level of cell carbon. Consequently, when growth is limited by a substrate in this category, cell growth is energy-limited. Furthermore, the maximum cellular yield is set inherently by the ratio between the biologically available energy and the carbon content of the organic molecule. The limiting factor of the growth yield with substrates having heats of combustion higher than 11 kcal/g substrate carbon is unknown, but it is unlikely that the energy potential from the substrate is the limiting factor, though there may be some examples of growth which are limited by the rate of supply of reductant (e. g., methylotrophs grown on methanol, see Anthony [18]).

At this point, it can be concluded that theoretical calculations provide only a preliminary, though valuable, basis for an estimation of maximal yields on different substrates. As was pointed out by Van Dijken and Harder [531], "a complicated series of events, resulting in the growth of living organisms, is very difficult to summarize in a few chemical reactions. It is the experimental yield, which may, or may not, be different from the calculated yield, that gives the final answer to the efficiency of growth of microorganisms on different compounds."

2.2.3.2 *Experimental Cellular Yields*

Table 19 shows experimental results of growth parameters obtained from steady-state cultures of different microorganisms grown on various substrates under carbon-limitation conditions. Three main conclusions can be drawn from the data.

1) The less oxygen in the carbon substrate, the higher the yield coefficient that is obtained. This is shown from yields obtained for growth of a given microorganism (i. e., *Pseudomonas* 1 or *Pseudomonas* 135) on methanol, formaldehyde, and formic acid; the last substrate resulted in the lowest yield. The high yield of cells growing on a relatively reduced carbon substrate, e. g., methanol as compared to formic acid, is counteracted by the greater oxygen requirements of these cells, and the heat evolved per unit of SCP produced (see below).

2) The biochemical pathways for assimilation of the carbon substrate determine the yield obtained. This is evident from the higher experimental yield values obtained for growth on methanol of RMP-pathway bacteria (*Pseudomonas* C and *M. methylotrophus*) as compared with serine-pathway bacteria (*Pseudomonas* 1 and *Pseudomonas* 135). In bacteria which utilize methanol via the RMP pathway, the efficiency of carbon assimilation is between 59 to 65% while in bacteria which possess the serine pathway, lower values were obtained (38.5 to 47.5%) [346]. Also, the values for maximal growth rate (μ_{max}) of the RMP-pathway bacteria are much higher (0.49 to 0.69 h^{-1}) than those for the serine-pathway bacteria (0.069 to 0.24 h^{-1}) [346].

3) The yields obtained for yeasts are lower than those for bacteria. This conclusion is evident by comparing the yields of *S. cerevisiae* and *E. coli* on glucose, *H. polymorpha* and *P. pastoris* with different RMP-pathway bacteria on methanol, and *S. cerevisiae,* and *C. utilis* with *A. calcoaceticus* for growth on ethanol and acetic acid, respectively.

2.2.3.3 Maintenance Energy Requirements and "True Growth Yield" Values

The maintenance energy of the microorganism is among the factors that play a role in determining the yield of cells obtained per unit of carbon and energy source. This term, first suggested by Pirt [539], may be defined as that portion of the biologically available energy, produced by catabolism, which is not available for biosynthetic purposes. It represents utilization of ATP for maintenance of concentration gradients between the cell and the medium, for motility, for preservation of the right ionic concentration, for turnover of cellular constituents, and wastage of ATP by hydrolysis. The extra substrate consumption required for such maintenance catabolism, therefore, would affect growth yields so that they become growth-rate dependent [539].

Although several models for estimating maintenance energy have been developed [519, 530, 540–542], it is convenient to characterize it in terms of a maintenance coefficient, m_s, which expresses the specific rate of consumption of substrate for maintenance. This coefficient can be determined by measuring the yield at different dilution rates in the chemostat and plotting $1/Y_s$ vs. $1/\mu$ (or $1/D$) to solve the Eq. [459]:

$$1/Y_s = 1/Y_s^{max} + m_s/\mu, \qquad (9)$$

where Y_s is the observed yield (g cell dry wt/g substrate consumed), Y_s^{max} is the "true growth yield" expected when maintenance is zero (see below), μ is the specific growth rate (h^{-1}) equal to the dilution rate (D, h^{-1}) at steady state, and m_s is the maintenance coefficient (g substrate/g cell dry wt·h). The slope (m_s) and intercept ($1/Y_s^{max}$) can be calculated from these plots.

Table 19 presents results of m_s values obtained for different organisms grown on various carbon substrates. Although the maintenance coefficents are extremely variable and range from 0.004 to 0.447 (g substrate/g cell dry wt·h) for various organisms growing on different substrates, it appears that the m_s values of yeasts are lower than those of bacteria (compare values for growth on ethanol, acetic acid, glucose and methanol), and that the m_s value does not appear to vary systematically with the carbon substrate (compare values for *Pseudomonas* 1 grown on methanol, formaldehyde and formic acid) (Table 19). The maintenance coefficients of bacteria growing on methanol, unlike their yields, appear to be similar and independent of the metabolic pathway used (Table 19).

We have found that growth conditions affect the maintenance requirements of *Pseudomonas* C during growth on methanol [348, 526]. Under carbon and energy limitation conditions, the m_s value was 0.072 and the Y_s value was 0.54. With carbon limitation and an excess of energy $m_s = 0.047$ and $Y_s = 0.6$, while during growth under excess of carbon and energy (nitrogen limitation) $m_s = 0.157$ and $Y_s = 0.24$. These results clearly show that for SCP production there is an incentive to employ conditions under which the growth of organisms is limited by carbon but not by energy.

It is worth mentioning that maintenance coefficients for mixed cultures were reported to be about 10 times lower than for axenic cultures [543]. As the specific growth rates (residence times) of the mixed cultures, i.e., activated sludge, were lower than those of the axenic cultures, it is conceivable that at low dilution rates a selection occurred for species with very low maintenance requirements. Such a selection was also possible in the case of methanol-grown bacteria. A *Pseudomonas* sp. was isolated and grown on methanol, and had a very low m_s value (0.02 g methanol/g cell · h [544], as compared with values ranging from 0.048 to 0.125 for other methanol-grown bacteria (Table 19)).

As will be discussed later, the kinetic parameter m_s should be considered together with the relevant yield factors of carbon source (Y_s), oxygen (Y_o), and energy input (Y_{kcal}), when searching for a suitable microorganism, raw material and cultivation methods for large-scale SCP production. For this process, it is of importance to find out the minimal maintenance requirements of the system and to apply growth conditions under which no more than this minium is diverted for nongrowth purposes, thus approaching the value of "true yield."

The "true yield" or maximum cell yield (Y_s^{max}) represents the yield which would be obtained if maintenance were reduced to zero. Table 19 presents the experimental "true yield" values obtained for different microorganisms and carbon sources.

It should be noted here that the higher yield values of the RMP-pathway bacteria are not connected to their maintenance requirement ($m_s = 0$), but presumably reflect either the lower ATP requirement per unit of biomass produced or possibly the more efficient production of ATP per unit of substrate. The ratio between the Y_s^{max} values of the RMP-pathway bacteria and those of the serine-pathway bacteria found experimentally was about 1.4 (Table 19) [526].

2.2.3.4 *Productivity*

An important question for the basic concept of an SCP production process is how to gain optimal productivity from a given type of fermentor. The limits of the productivity are determined by the maximal specific growth rate and the cell concentration. When a given bacterium has a high μ_{max} value, then the limit of productivity is related to the ability of the equipment to effect oxygen transfer and remove heat, i.e., the maximum limit of the dilution rate is occasionally determined by equipment limitations. According to Solomons [8], cell biomass dry weights can be obtained within the range of 15 to 30 g/liter, and dilution rates of 0.1 to 0.5 h^{-1}, providing productivities of 1.5 to 15 g/liter · h. However, due to problems of availability of dissolved oxygen, and/or heat removal, productivity values in the lower range of 3 to 5 g/liter · h were more common, though much higher values (up to 25 g/liter · h) effected in small-scale culture vessels, were reported.

The overall productivity of a production plant is also dependent on the percentage of the working year that the plant is operative. Although most SCP processes are based on continuous operation and hence are expected to operate 8000 h per annum, actual operativities data, though rarely published, are much lower. Solomons [8] quotes yearly operation runs of 1800 h (The Pure Culture Products), 2000 h ("Pekilo process"), and 2500 h (ICI), with anticipations of the ICI team to achieve continuous aseptic operation runs of 4000 h and possibly 8000 h. It can thus be estimated that a plant utilizing a 1000-m^3 fermentor with a productivity value of 5 g/liter·h and operation runs of 2000 h/y can produce 10,000 ton/y of SCP product, containing about 5000 ton/y of protein.

It should be noted that it takes considerable time to prepare media, sterilize vessels and reach steady state in continuous culture. Thus, the concept of "downtime," or plant inoperativity exists for continuous processes too. The effects of contamination during a continuous process, planned to run for several thousand hours are:

- loss of product in the fermentor and at various stages of processing,
- necessity to stop continuity of the process,
- resterilization of media and hardware,
- new inoculation,
- attainment of steady state once more, and
- the disposal of a large volume of high BOD material.

Frequency of contamination can thus determine the economic viability of a continuous process [36].

2.2.3.5 *Relationship among Cell Yield, Oxygen Demand, and Heat Load*

The relationship between yield based on the carbon source and the oxygen requirement (Y_o-yield coefficient of oxygen in g cells/g O_2) was calculated on the assumption that the products of metabolism are cells, water and carbon dioxide [545]. In addition, it was assumed that an inorganic nitrogen source is employed for growth. Both assumptions are usually valid for SCP production. The generalized equation for calculating the oxygen yield coefficient was shown by Mateles [545] to be:

$$\frac{1}{Y_0} = \frac{gO_2}{g\,\text{cells}} = 16\left[\frac{2C+H/2-O}{Y_sM} + \frac{O'}{1600} - \frac{C'}{600} + \frac{N'}{933} - \frac{H'}{200}\right]. \qquad (10)$$

An approximation would give

$$\frac{1}{Y_0} = \frac{32C+8H-16O}{Y_s \cdot M} - 1.58\,, \qquad (11)$$

where C, H, and O represent the number of atoms of carbon, hydrogen and oxygen, in the substrate, respectively. Y_s is the substrate yield coefficient in g cells/g substrate, C′, H′, O′, and N′ represent the percentage of carbon, hydrogen, oxygen, and nitrogen in the cells respectively and M represents the molecular weight of the carbon source. Using the latter Eq. and published data on the cell composition of bacteria and yeasts, the oxygen yields for various types of sub-

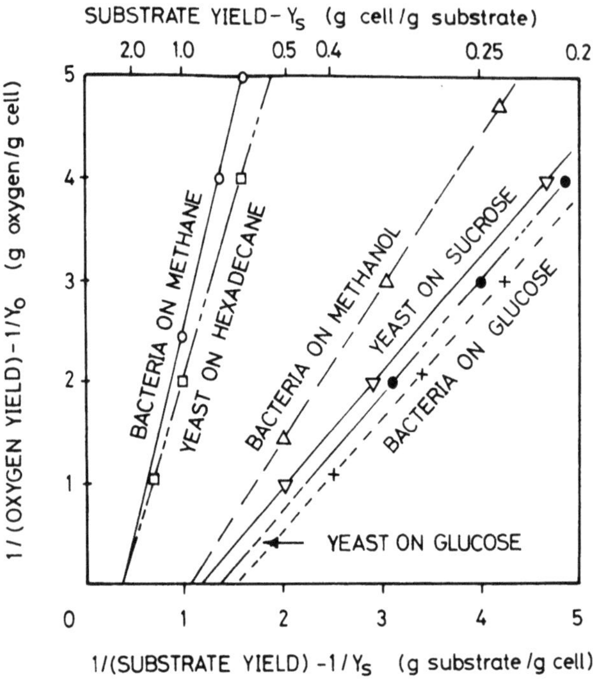

Fig. 23. The relation between substrate yields and oxygen yields for different microorganisms grown on different substrates. (Redrawn from Wang et al. [463]). (Reprinted with permission of John Wiley & Sons, Inc.)

strates were calculated by Mateles [545] and are shown in Fig. 23. It can be seen that the substrate and oxygen yield coefficients are inversely related. Therefore, the higher yield obtained by replacing carbohydrate substrates with hydrocarbon substrates (hexadecane and methane) was counteracted by the greater oxygen requirements of these cells per unit of SCP produced.

During active aerobic growth when the maintenance requirement is low, heat evolution is a result of electron transfer to atoms of oxygen and is therefore directly proportional to oxygen consumption rate or to the rate of cell mass synthesis. The heat produced during the course of fermentation can be estimated quite accurately by using the quotient 120 kcal per mole O_2 taken up, a figure which was obtained experimentally by Cooney et al. [546] using different substrates and microorganisms. Therefore, it can be concluded that a low amount of heat evolved per unit of biomass produced is obtained when the conversion efficiency of carbon substrate to cell carbon is high, i.e., the heat evolved (Y_{kcal}; g cells/kcal evolved) is inversely related to the cellular yield. Again, another unwanted result of the higher yield on hydrocarbon substrates, as compared with those on carbohydrates, is the higher heat production on hydrocarbons.

The yield coefficients, Y_s, Y_o, and Y_{kcal}, are dependent, not only on the carbon source and the microorganism, on the specific growth rate of the microorganism and its maintenance coefficient. The latter two interactions were evaluated in detail by Abbott and Clamen [547]. Their analysis disclosed that large increases occurred in all three yield coefficients as the growth rate of the bacterium (*Pseudomonas fluorescens*) was increased from 10% to 100% of μ_{max}; the relationship between μ and Y_o or Y_{kcal} was linear while Y_s exhibited nonlinear behavior. When cost factors were assigned to the various yield coefficients, operating cost was almost constant for growth rates between 50% and 100% of μ_{max}, but it rose rapidly when growth rates declined below 50% of μ_{max}. The rates at which the yield coefficients changed as μ was changed were dependent on the value of the maintenance coefficient (where μ was constant at 50% μ_{max}). When calculations were made applying different growth rates, it was found that the dependence of yield coefficient on maintenance coefficient was greatest at lowest growth rates, and that there is a strong economic incentive to employ a microorganism with a low maintenance coefficient for SCP production [547].

2.2.4 Physiological Problems Associated with Scale-up

The scale-up of fermentations and fermentors is among the more critical problems associated with fermentation technology, mainly because an important nutrient, oxygen, is required in large quantity by the cells, and cannot be dissolved to a high enough concentration in the nutrient medium. Due to the larger height of a production bioreactor as compared with a small scale reactor, there is an increase in the hydraulic pressure, resulting in higher partial pressures of oxygen and carbon dioxide, effecting changes in the growth conditions. While in the case of a small laboratory fermentor it is relatively easy to secure adequate mixing of the contents because no gradients exist within the liquid phase, in a large fermentor there are likely to be gradients caused by non-ideal mixing. Some of these gradients relate to the concentrations of dissolved oxygen and carbon dioxide, or to the limiting nutrient, when the fermentor is operated as a chemostat. It is important to know how these time-varying parameters affect the fermentation process [36, 548]. Other constraints that exist in scaling-up fermentation and fermentors are the rheological properties of fermentation broths and heat transfer considerations, as discussed elsewhere [517, 548, 549].

2.2.4.1 The Effect of Oxygen

Single cell protein processes are aerobic and therefore have a high demand for a continuous oxygen supply. Taking into account the low solubility of oxygen in the medium (about 7 mg/liter for broths in equilibrium with air at 1 bar and 20 °C) [550], and the size of a production fermentor, it becomes increasingly difficult to fulfill the demand for oxygen (between 2 and 10 kg $O_2/m^3 \cdot h$) in scaling-up of aerobic fermentations [548]. In response to low dissolved oxygen tensions (DOT $<$ 0.08 atm) *Methylomonas methanolica* WLO [551] and *Methylomonas flagellata* [552], showed a very sharp decrease in cell yield. A lowering of the yield coefficient was also observed when *M. methanolica* was exposed to low DOT for short periods of time at regular intervals. The results are comparable to those obtained for *Pseudomonas* AM-1 where a decrease in cell yield at low DOT was accompanied by an increase in respiration rate [553–556]. However, the "critical" DOT reported for *Pseudomonas* AM-1, at a cell concentration of about 5 g/liter, is lower than the value found for *M. methanolica*. *Pseudomonas* AM-1 also appears to be affected more than *M. methanolica* by high DOT tensions, and have similar responses to those obtained with *Methylomonas clara* [557, 558]. In a continuous culture, there was no influence of oxygen (in the case of oxygen-unlimited growth) on the yield up to a partial pressure of 0.7 atm. Using pure oxygen for aeration (1 atm), the yield decreased by 35% of the average value. The toxic effect of oxygen can be ascribed mainly to superoxide formation.

The behavior of a steady state aerobic culture of *Hansenula polymorpha* to an interruption in the oxygen supply was simulated by a three-hour reduction in the rate of agitation of methanol-grown cells in a chemostat [559, 560]. Immediately following agitation reduction, the residual methanol concentration began to accumulate and the cell concentration began to fall. A methanol concentration of approximately 12 g/liter was reached before agitation was resumed. When oxygen again became available there was a very rapid accumulation of formaldehyde and formate as intermediate products, both of which are extremely toxic to *H. polymorpha*. These results coincide with those reported by Pilat and Prokop [561] for *Candida boidinii* Bh 11. With the use of computer control strategies, it was possible to avoid the transient increase in methanol concentration by controlling oxygen supply, and, as a consequence, prevent production of formaldehyde and formate [560].

2.2.4.2 The Effect of Carbon Dioxide

When *Pseudomonas* C was grown on methanol, a yield value of 0.54 g cell dry wt per g methanol was obtained, with the conversion of 65% of methanol carbon to cell carbon. For each mole of oxygen consumed, 0.5 mole of carbon dioxide was produced [346]. If we assume that the aeration rate is such that 80% of the oxygen is utilized, then the composition of the gas stream leaving the fermentor is: $pO_2=0.046$ atm, $pCO_2=0.092$ atm and $pN_2=0.86$ atm [36]. Thus, the gas stream is more or less stripped of its oxygen and is fairly rich in carbon dioxide [36]. It is clear that in a large fermentor there is an increase in the total pressure resulting in an increase in the partial pressures of CO_2, in both the gas and the broth, and this must affect the growth rate. In the case of the ICI's pressure cycle fermentor, a fermentor height of 42 m is involved; a pressure of 5 atm is produced. Only a few reports have been published on the effect of CO concentration on biomass production. It's inhibitory effects on growth are reversible. In *Saccharomyces cerevisiae* 50% carbon dioxide in the aeration mixture caused a 16% decrease in cell yield [561]. A decrease in the yield coefficient of *Methylomonas methanolica* was observed at a carbon dioxide partial pressure of 0.3 atm and above [551]. At 0.6 atm the yield coefficient was only 70% of the maximum value observed (0.48 g cell dry wt per g methanol utilized). Carbon dioxide also had an inhibitory effect on the growth of *Methylomonas clara* on methanol [557, 558]. Depending on the dilution rate, inhibition started at 0.3 atm, and at 0.5 atm the yield was reduced to 50% of standard values. Inhibition seemed to be a diffusion problem of CO_2 from the cells to the liquid phase, rather than a regulatory effect, as inhibition was observed in all cases to be reversed within one minute.

Many microorganisms require small amounts of carbon dioxide for rapid growth [459]. Under the growth conditions normally contemplated for SCP production, the concentration of carbon dioxide is certainly adequate. The minimum partial pressure of carbon dioxide necessary to avoid reductions in yield for the case of the RMP-pathway bacterium *Pseudomonas* C growing on methanol was estimated to be 0.003 atm, or about ten-fold higher than its normal concentration in the atmosphere [265, 563]. Exceptionally high yields that were obtained in experiments with *Pseudomonas* TB582 at low methanol concentrations together with isotope and enzymatic studies, showed that CO_2 fixation enhanced the efficiency of methanol utilization [564]. These authors have suggested that by operating a continuous culture fermentor around the critical level of 0.26% methanol feed, one effects major advantages, such as a very low effluent substrate loss, and a 45% increase in yield, due to atmospheric CO_2 feeding. However, it should be noted that the effect of CO_2 on the yield is obtained only at low concentrations of cells, whereas in denser cultures this effect is not pronounced.

2.2.4.3 *Variations in the Concentration of Limiting Nutrients*

All submerged microbial cultures are subject to perturbations in their environment, due to imperfections in mixing. In small laboratory fermentors the energy input per unit volume is uneconomically large and perturbations are usually quite small, whereas in a large-scale fermentor perturbations are large. Since large-scale SCP processes are operated as continuous cultures with the carbon source as the limiting nutrient and with a small excess of oxygen, there will be variations in the concentration of the limiting nutrient in different parts of the fermentor. These variations are significant to the physiology of the cells. Here, examples of two carbon sources, glucose and methanol, will be discussed (for more details see Harrison and Topiwala [484]).

Hansford and Humphrey [565] have reported that in a small (10 liter) continuous fermentor operating at a low dilution rate, the yield of *Saccharomyces cerevisiae* increased by 12%, when the number of medium entry points was increased from one to three. Ryu [566] simulated the time-dependent effects of poor mixing by using a continuous culture which, instead of being fed continuously, was fed in "shots" at varying frequencies; the average feeding rate, and hence the dilution rate, was held constant. Using *S. cerevisiae* with glucose as the limiting nutrient, no reduction in yield was observed at dilution rates of 0.01 to 0.034 h^{-1} when the interval between shots was as long as 18 min. Similarly, using glucose-limited cultures of *Arthrobacter globiformis*, Luscombe [567] found no reduction in yield when using different shot intervals (time between two additions) at different dilution rates.

When methanol was supplied either discontinuously or in a concentrated feed, a substantial drop in cell yield was obtained. The feeding of a concentrated methanol solution separate from the other medium components, resulted in a dramatic decrease in the cellular yield of *Methylomonas methanolica* WLO, which showed that the organism was sensitive to local high methanol concentrations in the broth

[551]. Discontinuous addition at very short intervals (10 s) of the whole substrate flow, including methanol, decreased the biomass yield. When the pulse time was increased to 50 s, the yield coefficient decreased to 0.39 from a maximum value of 0.48 (g cell dry wt/g methanol utilized), obtained from a continuous flow of medium containing methanol. These results imply that continuous feeding of methanol with other medium components and a good distribution of the medium feed are important in reducing nonhomogeneous conditions, preventing gradient development in large-scale fermentors, and hence poor yields.

Analysis of the transient period following a pulse addition of methanol showed a transient accumulation of formate in the broth; indicating that in the energy metabolism of *M. methanolica* WLO, the NADH-yielding step (formate dehydrogenase) was not operating optimally. This resulted in more methanol being used for energy production, increased the probability of NADH limitation, and hence decreased the cellular yield [551]. Similar results were obtained with *Methylophilus methylotrophus* growing on methanol as a sole source for carbon and energy [568]. With the latter organism, laboratory fermentors gave better average yields than the pilot plant fermentors; the yield depended on the fermentor configuration [569].

The solution to the problem of nonhomogeneity in methanol concentration, implemented in the ICI single cell protein production plant, was to introduce methanol through many small nozzles throughout the fermentor. This was a complicated and expensive solution whereby methanol was added through 3000 nozzles in the ICI 1500-m^3 commercial fermentor, resulting in a steady state methanol concentration in the fermentation broth of 2 ppm [569].

Methanol-grown yeasts are probably more resistant than bacteria to changes in methanol concentrations in the culture broth. When the effect of discontinuous methanol feeding on *Hansenula polymorpha* DL-1 was examined, it was found that culture stability was not affected, and in no case did a significant decrease in cell concentration or increase in methanol concentration result from the oscillatory feeds [560].

2.2.5 Sterilization

One major problem which ICI faced in satisfactorily operating continuous production of "Pruteen", was to maintain sterility [570]. Only by "meticulous attention" to equipment, systems and procedures, and a considerable investment, have they been able to achieve long running periods. Obviously, the full economic potential of a continuous fermentation is realized in long production runs. This is only possible if the SCP process operates with the inoculated microorganism present in the fermentor, without any contaminating microorganisms. To perform this operation, systems to maintain monoseptic conditions are required (for details see Wang et al. [463] and Sikyta [486]). These systems include three basic phases relating to the operation of the plant [571]. Firstly the fermentor and attached piping and valves are sterilized by steam (presterilization). Secondly, the various streams fed into the fermentor either before or during the process are sterilized by heat and/or filtration. Sometimes no sterilization is required because the inherent microbicidal properties of the stream suffice. Thirdly, sterile barriers at the interface between the fermentor and the outside environment must be maintained. This is usually done by steamlocks involving a pair of valves between which live stream is maintained.

An SCP process should be subject to less contamination owing to operator error, i.e., opening the wrong valve or failing to use proper aseptic procedures in taking samples, because in continuous processes there is less need for operator intervention than in batch fermentations [36]. Still, the complexity of the sequences involved in operating process connections and control instrumentation is often greater than operators can manage and, therefore, micro-processors are used to coordinate sequences and to provide saftey interlocks in order to prevent contamination; e.g., two computers for ICI's plant [571].

2.2.6 "Structured" Mixed Cultures

In spite of the dependence of so many traditional processes on the cooperative action of various microbial species, it is undeniable that the modern fermentation industry owes its existence to the development of monoculture techniques [572]. Nevertheless, mixed culture systems are the norm in nature and their use for production of food is by no means novel; traditional foods, such as sauerkraut, yogurt, cheese and soy sauce have been consumed by man for thousands of years. It was commonly held that to obtain the consistency of control required for the acceptance of any new food or animal feed product, it would be necessary to use monoculture systems. This prejudice was implicit in the International Union of Pure and Applied Chemistry (IUPAC) guidelines (see also the Protein Advisory Group of the United Nations University – PAG/UNU guideline No. 12, 1980) [573], which stated that cultures used to produce SCP should be clearly identified. Harrison and his group at Shell Laboratories in England argued that these strict guidelines in no way preclude the use of mixed cultures for SCP production, since there was no reason that these cultures could not be defined in terms of the component microorganisms. They used the term "structured mixed cultures" to denote those cultures that were obtained by recombination of defined microorganisms, rather than by enrichments yielding undefined and uncontrolled mixtures of microorganisms [572].

Although Harrison recognized that "microbiologists have neglected the study of mixed microbial cultures, regarding them as experimental nightmares", he pointed out that the major incentive for developing mixed culture technology for SCP production evolved when people began considering complex carbon sources as raw materials for biomass production [556, 572].

2.2.6.1 Complex and Multiple Carbon Sources

Most microorganisms utilize soluble sugar much more rapidly than insoluble carbohydrates and, therefore, the cellulosic material of the biomass must be hydrolyzed before it can be fermented to SCP or to other desired end products. It can be expected that when lignocellulosic waste materials are used for SCP production, the reactor will be subject to a continuously changing feed. This is caused by variable flows and fluctuations with respect to both the nature of the organic compounds and their individual concentrations, together with imbalances in the respective organic and inorganic loading. In addition, when microorganisms are grown on multisubstrates, such as wood hydrolyzate, a phenomenon of diauxic growth and sequential substrate utilization is often encountered [446, 574]. Mixed cultures are used for the production of SCP from this complex natural substance mainly because it is not readily attacked by one microbial species. A mixed culture containing two yeast species and four bacterial species was described [401]. This system degraded calcium lignosulphonate at a much higher rate than was observed with various pure cultures [Table 20].

Starchy materials, such as potato and rice processing wastes, must be converted to mono- and disaccharides to be suitable as substrates for SCP produc-

Table 20. Examples of "structured" mixed cultures and their advantages

Raw material	Organisms	Benefits of culture	Ref.
Methane	A methane-utilizer + a methanol-utilizer + a citrate-utilizing bacterium	High stability	[575, 576]
Methane	*Pseudomonas* sp. + *Hyphomicrobium* sp. + *Flavobacterium* sp. + *Acinetobacter* sp.	Removal of the toxic methanol-inhibiting growth of *Pseudomonas* sp. and removal of organic byproducts resulting in high stability	[345, 577]
Methane	*Methylococcus* sp. NCIB 11083 + *Flavobacterium* sp. NCIB 11282 + *Pseudomonas* sp. NCIB 11309 + *Pseudomonas* sp. NCIB 11310 + *Moraxella* sp. NCIB 11308 + *Nocardia* sp. NCIB 11307	Culture was stable under methane limitation and high concentration of ethane	[578]
Methane	*Methylococcus* sp. + *Pseudomonas* sp. NCIB 11310 + *Mycobacterium rhodochrous* NCIB 11307 + *Moraxella* sp. NCIB 11038 + *Pseudomonas* sp. NCIB 11309	High stability, lack of foaming and growth at 45 °C	[556, 579–580]
Methane	*Methylococcus ucrainicus* + *Rhodopseudomonas glutinis* + *Methylomonas methanica* + *Pseudomonas fluorescens*	Less inhibition by added amino acids	[581]
Methanol	*Pseudomonas* sp. EN + *Pseudomonas* sp. NCIB 11019 + *Pseudomonas* sp. NCIB 11022 + *Acinetobacter* sp. NCIB 11020 + *Curtobacterium* sp. NCIB 11021	High stability, lack of foaming and optimal growth at 42 °C	[582]
Methanol	*Methylomonas methylovora* + *Xanthomonas* sp. + *Flavobacter* sp. + *Pseudomonas pseudoalcaligens*	High stability and no requirement for yeast extract	[583]

Table 20 (continued)

Raw material	Organisms	Benefits of culture	Ref.
Methanol and ethanol	Biotin-requiring *Hansenula polymorpha* M-2 + vitamin prototroph *Candida krusei* E-10	No vitamins are required in the medium	[584]
Glucose	*Aeromonas* sp. + acetate-utilizing bacterium	Removal of acetate which inhibits the growth of *Aeromonas* sp.	[585]
Date carbohydrate (Glucose and fructose)	*Candida utilis* + *Saccharomyces cerevisiae*	Better carbon conversion from substrate to SCP	[141]
Agave juice (carbohydrate)	*Saccharomyces carbajali* + *Candida utilis* *Saccharomyces carbajali* + *Candida parapsilosis*	Higher protein yields and higher lysine content	[75]
Lactose	*Streptococcus cremoris* + *Pseudomonas stutzeri*	Increased molar yield of *S. cremoris* on lactose due to lactate consumption by *P. stutzeri*	[586]
Sulfite waste liquor	*Candida utilis* + *Mycotorula* sp.	Better sugar consumption and better growth yield of *Mycotorula* sp.	[587]
Potato waste ("Symba" process	*Saccharomyces fibuliger* + *Candida utilis*	*S. fibuliger* hydrolyzes starch enabling the rapid growth of *C. utilis*	[73]
Calcium lignosulphonate	*Trichosporon cutaneum* + *Trichosporon fermentans* + *Arthrobacter tevregens* + *Arthrobacter citreus* + *Pseudomonas putida* + *Chromobacterium violaceum*	Higher degradation rate of raw material as compared to monoculture	[89, 401]
Barley straw	*Cellulomonas* sp. + *Candida utilis*	Better productivity and stability; no need to add yeast extract	[455]
Ryegrass straw	*Cellulomonas* sp. ATCC 21399 + *Alcaligenes faecalis*	Six fold increase in biomass yield	[451]
Wheat straw	*Paecilomyces fusisporus* + *Trichoderma viride*	Higher competitive ability for growth	[454]
Wheat straw	*Chaetomium cellulolyticum* or *Trichoderma lignorum* + *Candida lipolytica*	Increased substrate utilization and biomass production	[453]
Bagasse	*Cellulomonas* sp. + *Alcaligenes faecalis* (or yeast)	Higher protein production	[449, 450]
CO_2	*Chorella* strain 211/8k + 3 heterotrophic bacteria (mixed culture MA003)	A most efficient convertor of light energy into biomass. Lower maintenance coefficient	[167]

tion. The Swedish Sugar Corporation developed the "Symba" process for treatment of wastes that contain starch. The two organisms used are *Saccharomyces fibuliger*, which produces α- and β-amylases for hydrolysis of starch to glucose and maltose, and *Candida utilis* which utilizes these sugars as a substrate for growth (Table 20) [395].

Natural gas is found either by itself or in association with oil. The main component of natural gas is methane, and considerable quantities of ethane, propane, butane and pentane may be present. Malashenko et al. [588] examined 147 bacterial isolates growing on natural gas, and showed that some were capable of growth on either C_1 or C_2 to C_4 hydrocarbons, but none was capable of simultaneously utilizing C_1 and C_2 to C_4 gaseous hydrocarbons. Romanovskaya et al. [589] found that ethane and propane were competitive inhibitors of methane oxidation. Therefore, in order to utilize the multiple-carbon substrate efficiently and to overcome the inhibition of growth of the predominant methane-utilizer by toxic higher alkanes, a mixed culture is obviously preferable to monoculture. A mathematical model was developed to predict the performance of a mixed culture grown on natural gas. A high productivity and stable culture growing under methane limitation with high ethane and propane concentrations was achieved (Table 20) [590, 591].

Another example of a multiple-carbon substrate was studied by Wilkinson and Hamer [592] using an artificial effluent, containing methanol, phenol, acetone and isopropanol. It was shown that the use of a mixed culture was efficient. This system behaved as if it consisted of three superimposed non-interacting components responsible for oxidation of methanol separately, phenol separately, and acetone together with isopropanol.

Heavy gas-oil is a mixture of hydrocarbons containing n-paraffins, isoalkanes, alkenes, cycloalkanes and aromatics. The composition of gas-oil varies in different batches. In the gas-oil-SCP process developed by British Petroleum utilizing a monoculture of *Candida tropicalis*, about 90% of the hydrocarbon feedstock remained unused after the fermentation [63, 593]. Besides the inefficiency of carbon utilization, the residual carbon had to be removed from the yeast cells, recovered, and then returned for use in normal refinery processes. The yeast cells in turn had to be washed, and extracted with solvent to minimize any residual gas-oil. These treatments led to a substantial increase in the price of the SCP product.

A mixed culture containing the bacterial strains RAG-1, UP-2, UP-3, and UP-4 was described by Horowitz et al. [594]. When these four strains were grown together on gas-oil, there was over 75% oil conversion. By plating out on selective media, growth of the individual strains was measured kinetically in the reconstituted mixed culture, revealing competition for common growth substances of UP-2 and RAG-1. Increased die-off of UP-2 and stabilization of UP-4 were observed during the stationary phase.

It must be pointed out that mixed cultures are not the only solution for efficient utilization of a multiple-carbon source. Horowitz et al. [594] have suggested the sequential growth of different microorganisms on a complex (or multiple) substrate. Thus, when crude oil that had been recovered after exhaustive growth of either RAG-1 or UP-2 was used as a carbon source for further growth of the strains UP-3 and UP-4, over 75% oil conversion was measured; the result is similar to that obtained with a mixed culture composed of these strains.

The other alternative is the construction of a bacterium that is able to metabolize most of the major hydrocarbon components of petroleum. By means of successive crosses a "superbug" was devised by Chakrabatry that carried the XYL plasmid that is responsible for both xylene and toluene utilization and NAH plasmid that is responsible for naphthalene utilization. The bacterium contained also a hybrid plasmid derived by recombining parts of CAM (utilization of camphor) and OCT (utilization of octane, hexane and decane). The multiplasmid-containing bacterium (*Pseudomonas* sp.) grew rapidly on crude oil because it metabolized much more of the hydrocarbon components than any other common microorganism [595].

2.2.6.2 Defined and Simple Carbon Sources

In cases where single cell protein production employs a single, defined and simple carbon source such as sugar, ethanol, methanol, methane, CO_2, etc., we can identify three main classes of mixed cultures of microorganisms which might exist. These are: 1) cultures in which the component microorganisms compete for the primary substrate, 2) cultures in which one component microorganism grows on the primary substrate, producing organic compounds which serve as substrates for other components of the system, and 3) cultures in which there is a complex interaction of the component microorganisms owing to mutual supplementation of nutritional deficiences or mutual antagonism [596].

Competition for the Primary Carbon Substrate

Mixed cultures of this type were shown to be unstable, in that one or another of the component microorganisms will rapidly dominate virtually the entire continuous culture leading to the disappearance of the other component (unless one component exhibits wall attachment, i.e., the environment becomes inhomogeneous – see Baltzis and Fredrickson [597]). Thus, if the desired microorganism is the dominant one, there is no serious problem, whereas if the contaminant microorganism is the dominant one, the production run will come to a rapid end. (For more details, see the theoretical work of Aris and Humphrey [598] and the review of Slater and Bull [599]).

Competition among methylotrophs for the carbon source has been studied by Rokem et al. [600]. Competition for methanol as the sole carbon source between the obligate methanol-utilizer (RMP pathway), *Pseudomonas* C, and the facultative methylotrophs (serine pathway), *Pseudomonas* 1 and *Pseudomonas* 135, was observed. In spite of the lower efficiency of serine pathway bacteria to methanol, *Pseudomonas* 1 and *Pseudomonas* 135 outcompeted *Pseudomonas* C in the mixed culture, during methanol-limited growth, at a dilution rate of 0.1 h^{-1}, but at a dilution rate of 0.3 h^{-1}, *Pseudomonas* C prevailed, and the other bacteria were excluded from the chemostat. *Pseudomonas* C had a lower affinity constant (K_s) for methanol than the other bacteria. These results showed the importance of both K_s-value and μ_{max} for microbial growth on a single growth-limiting substrate. They also confirmed the apriori assumption that the yields of these mixed cultures can never be higher than those obtained by growing axenic cultures of the RMP-pathway bacterium at optimal conditions [265].

It is noteworthy that such behavior of component microorganisms is not restricted to mixed cultures in which the limiting nutrient is the organic carbon source, though in the case of SCP production it would be the preferred.

Comparative studies were performed on the competition for thiosulfate and oxygen as energy sources, and for iron by chemolithotrophic sulfur-oxidizing bacteria. These established the principle that, depending on the fluctuation of the growth-limiting substrate concentration, or changes in environmental conditions (such as changes in pH), the component microorganisms may each possess an alternative ecological advantage. This explains the domination of one organism over the other in the culture [601, 602].

An interesting system can be seen in a chemostat culture in which species of *Chlorella* and *Spirulina* competed [46]. Without biomass recycle, the *Chlorella* displaced the *Spirulina* sp. When a portion of the harvested biomass, which preferentially concentrates the *Spirulina* sp., was returned to the chemostat, the retention time ($1/\mu$) of the desired algae artificially increased and thus the algae obtained a competitive advantage. With 67% of algal recycle, the *Spirulina* sp. displaced the *Chlorella* sp., demonstrating the potential of this technique for species control.

A different mixed culture system was described by Cremieux et al. [583]. The culture comprised one obligate methanol-utilizer (80% of population), *Methylomonas methylovora*, and three facultative methanol-utilizers (remaining 20% of population), *Xanthomonas* sp., *Flavobacter* sp. and *Pseudomonas pseudoalcaligenes*. This culture would obviously not be stable if the different component bacteria were competing only for the methanol. However, the mixed culture was stable and its composition remained unaltered for two months of continuous growth at a dilution rate of $0.3\ h^{-1}$. It is most improbable that all four organisms utilized methanol as their sole source of carbon and energy. Such a stable existence is only possible if other interactions occur within the mixed culture. It is evident that this was indeed the case because the facultative methylotrophs required yeast extract for growth in monocultures. Although the nature of the interactions involved in the mixed culture was not examined, the simplest explanation for its stability is that during growth on methanol the obligate methanol-utilizer excreted organic products (or that some lysis occurred), and the facultative methylotrophs subsequently grew on them (see the second class of mixed cultures). This interrelationship is analogous to that of the stable coexistence of the obligate *Thiobacillus neapolitanus* and the facultative chemolithotroph *Thiobacillus* sp. A2 during thiosulfate-limited growth in a chemostat [602, 603]. It was shown there that the obligate bacterium excreted substantial amounts of glycollate, which served as a carbon and energy source for the facultative bacterium.

Associated Cultures in Which One Component Microorganism Grows on the Primary Substrate, Producing Organic Byproducts Used by the Other Components

In any SCP fermentation process the substrate carbon should eventually terminate in the cell mass or in CO_2. Ideally, it is expected that other undesired bypro-

Table 21. Advantages of mixed cultures relative to monocultures

- Better utilization of a complex carbon source or of mixtures of carbon sources
- Removal of organic carbon excreted by one component microorganism:
  a) removal of toxic compounds affecting one component microorganism
  b) increase of the total biomass yield from the primary carbon source
  c) elimination of foam
  d) increase the growth rate of one component microorganism
  e) enable water recycling
  f) resistance to perturbations in culture environment
  g) resistance to contamination by bacteria, yeasts, fungi and phages
  h) lower maintenance energy requirement
- Attainment of desired protein compositions

ducts are to be minimized. The complete utilization and elimination of organic compounds from the culture broth by a monoculture, cultivated in a large-scale continuous process, have not yet been achieved. In all axenic (mono) cultures examined this far, the excreted carbon compounds comprise 2–10% of the steady-state bacterial biomass carbon [265, 580, 585, 604, 605]. The excretion of organic matter has several disadvantages: lowering of the biomass yield, accumulating inhibitory substances (e.g., methanol, formaldehyde, formate, acetate, amino acids, etc.,) and the possibility of being contaminated by bacteria which are not able to utilize the primary substrate. Furthermore, such excretion may limit the re-use of water after cell separation, as recycling organic compounds in the fermentor is undesirable; this will lead to their accumulation, and possibly introduce a toxicity factor into the system. The predominant disadvantage of excretion of organic matter is the possibility of culture contamination by undesired microorganisms, making control of the fermentation process difficult. One solution to this problem is to operate the SCP production process under strictly aseptic conditions [56, 57, 606]. However, it is only with high capital investment and operating costs that sterility might be maintained over a prolonged period of weeks or months in a continuous production plant [570]. The removal of organic byproducts from the fermentation broth overcomes the aforementioned problems. Hence, it was suggested that the SCP process be operated without taking aseptic precautions, besides general cleanliness, and to carry out the fermentation process with the second class of mixed cultures of microorganisms [596]. Mixed cultures of the second class offer certain advantages which were summarized in the comprehensive review by Linton and Drodz [578]. These are shown in Tables 20 and 21 and discussed below for growth on methanol and methane.

Methanol as a Sole Carbon Source. A mixed culture, established by means of continuous enrichment of a sample of mud, for the production of single cell protein from methanol was described by Harrison and his colleagues [556, 572, 582, 607, 608]. The five component bacteria of the mixed culture were isolated and identified. One of the component bacteria, *Pseudomonas* sp. strain EN, was an obligate methylotroph (able to grow only on methanol). The other four bacteria, which could grow on a variety of carbon substrates (heterotrophs) but not methanol, were identified as: *Pseudomonas* sp. NCIB 11019, *Pseudomonas* sp. NCIB 11022, *Acinetobacter* sp. NCIB 11020, and *Curtobacterium* sp. NCIB 11021. Experiments carried out in a "U"-shaped culture vessel consisting of two chambers separated by a dialysis membrane, and containing methanol medium, showed that growth of *Pseudomonas* sp. EN was optimal when all four heterotrophs were present in the opposite growth chamber. While *Pseudomonas* sp. EN was able to grow on methanol in one chamber, without the presence of the heterotrophic bacteria in the other chamber, growth of the latter organisms occurred only when the methanol-utilizing organism was grown in the opposite growth chamber. The use of ^{14}C-labeled methanol confirmed that approximetly 15% of the methanol carbon was excreted as organic products (protein, nucleic acids, lactic acid, pyruvic acid, etc.) by *Pseudomonas* sp. EN, and was utilized completely by the non-methanol-utilizing bacteria. The mixed culture constructed with these component bacteria was extremely stable and the relative numbers of the various species in the mixture did not alter during many hours of cultivation in a chemostat operating at different dilution rates.

At the first stage of their research, Harrison and his colleagues reported that the mixed culture performed better in terms of growth yield and maximum specific growth rate (μ_{max}) than did the mono culture of their methanol-utilizer, *Pseudomonas* strain EN. Growth yields of 0.3 to 0.54 g cell dry wt/g methanol utilized and μ_{max} values from 0.16 to 0.66 h^{-1} for the mono- and mixed-culture were observed [556, 608, 609]. This work was carried out in a medium containing citrate, and it was subsequently shown that citrate inhibited the growth of *Pseudomonas* sp. EN [610]. By omitting citrate from the medium, and optimizing the growth environment, strain EN performed almost as well as the mixed culture in terms of growth yield and growth rate [572]. The fact that no significant difference in yield is found between growth of the methanol-utilizer in a single-strain culture and growth in the mixed culture, is explained by the relatively small amount of methanol which is converted to other organic byproducts (excluding cells and CO_2). This, in turn, controls the small percentage of the methanol non-utilizing bacteria present in the mixed culture. In the case of the mixed culture described above, the relative numbers of methanol non-utilizing bacteria varied according to the dilution rate, i.e., between 0.8 and 7.6% of the total cell numbers [556]. Similar results were obtained when different heterotrophs were added to a culture of the methylotroph *Pseudomonas* C in the chemostat [600]. Changes in the proportions of the heterotrophic bacteria were noted in the culture, but their total number remained fairly constant throughout the fermentation (less than 5% of the total population). No difference in yield was found between growth of *Pseudomonas* C in a single-strain culture on an optimized medium [477] and growth in the mixed culture [600].

Thus it seems that mixed cultures of this type, with methanol as a carbon source, can occasionally give results that are better in terms of yields, growth rates or productivities, than can be obtained using the methanol-utilizing component bacterium, especially when the medium is not capable of supporting maximal growth of the main component organism. However, in the event of recycling of medium during the SCP production process, the concentration of the excreted organic matter in the medium would increase profoundly, and as a result the proportion of the associative methanol non-utilizers would increase to a level which would significantly contribute to the overall cellular yield.

Methane as a Sole Carbon Source. A methane-utilizing mixed culture was isolated by Hamer et al. [577] as a batch culture, and was later extensively studied in a continuous culture by Wilkinson et al. [345]. The mixed culture, which was extremely stable, contained four bacteria: *Pseudomonas* sp. – a methane-utilizing Gram-negative, rod-shaped organism, a *Hyphomicrobium* sp. – capable of growth on methanol but not on methane, and two Gram-negative rod-shaped organisms, incapable of growth on methane or methanol, but capable of growth on a wide range of complex organic compounds. These two bacteria were identified as an *Acinetobacter* sp. and a *Flavobacterium* sp.

On mineral medium agar plates, only the *Pseudomonas* sp. grew, when methane was supplied as the carbon substrate. However, upon transfer to liquid culture, this bacterium required the presence of the *Hyphomicrobium* sp. for sustaine growth on methane. Even so, such growth was poorer than that of the complete mixed culture, and it could be enhanced only when the other two heterotrophic bacteria were added. This mixed culture was an example of obligate symbiosis, and it was postulated that during growth on methane, the primary methane-utilizer produced trace amounts of methanol which inhibited its growth. The *Hyphomycrobium* sp. utilized the methanol and thus relieved the growth inhibition of the methane-oxidizer. The two other organisms, which were present at levels of approximately 1%, were believed to grow on organic complex products of growth or lysis of the primary utilizer which consisted of about 95% of the population.

The *Hyphomicrobium* sp. was ideally suited for its role in this relationship for two main reasons. The first is that the organism was able effectively to scavange methanol down to a level of a few micromoles, thus relieving inhibition of methane oxidation. This was because of its markedly higher affinity ($k_m = 8\ \mu M$), with respect to the primary methane-utilizing bacterium, for methanol, [611]. The second reason is its ability to use nitrate rather than oxygen as an electron acceptor, even at quite high oxygen tensions. As methane (or methanol) utilization is a highly aerobic process, the utilization of nitrate rather than oxygen at low dissolved oxygen tensions by *Hyphomicrobium* sp. would prevent competition for oxygen between this bacterium and *Pseudomonas* sp. [612].

Complex Interactions Between Component Microorganisms

In some cases, complex interactions exist between the different components of the mixed culture; materials excreted by one component organism are essential for optimal growth of the other culture component. Photosynthetic SCP production in high-rate algal ponds (HRAP) shows this kind of interaction. Under suitable conditions, the algae establish symbiosis with aerobic bacteria. Organic matter in the ponds is degraded by the aerobic bacteria, producing CO_2. In the HRAP the algae, which are CO_2-limited, will utilize this CO_2, producing biomass and oxygen, which will then be utilized by the bacteria for further degradation of organic matter [53, 175]. Similarly, a mixed culture MA003 consisting of *Chlorella* sp. and 3 heterotrophic bacteria was far more efficient at converting light energy into biomass energy, relative to a *Chlorella* monoculture [167]. Other interactions exist due to differences in absorption spectra. Hence, purple photosynthetic bacteria can be developed under a layer of algae. Photosynthesis in algae utilizes light waves mainly in the range of 700 nm, whereas purple bacteria perform photosynthesis by absorbing light waves in the range of 800–900 nm [172]. It was shown that the associative bacteria enhance the growth of algae and, in some cases, e. g., *Microcystis aeruginosa*, the algae failed to grow in the absence of the bacteria [613], It is feasible that these effects are due to production of vitamin B_{12} by the bacteria, but other causes may be possible.

Although it has been suggested that mixed cultures are more resistant to contamination than monocultures [582], the main problem with the use of mixed cultures in SCP production is the possibility of contamination by pathogenic microorganisms. This is not a question to which a definitive answer can be given, but it seems reasonable to expect that if tests are conducted on a mixed culture maintained over a period of months or even years, and if the results in terms of toxicological and, similarly, nutritional testing are good, then the SCP product should be acceptable for animal feeding [596].

Table 22. Concentration of bacteria in the broth and on the fermentor wall at the end of continuous fermentation[a]

Bacteria in the culture	Bacterial concentration	
	Broth (ml^{-1})	Wall (cm^{-2})
Pseudomonas C+	9×10^8	0.5×10^7
Soil bacteria	2×10^7	1.6×10^7
Pseudomonas C+	1×10^9	0.2×10^6
S. aureus	1×10^6	1.6×10^6
Pseudomonas C+	2.5×10^9	0.3×10^6
S. typhimurium	2.7×10^7	1.6×10^6
Pseudomonas C[b]+	1×10^9	0.3×10^7
Soil bacteria[b]+	1×10^8	3.7×10^7
S. aureus	$<1\times10^3$	$<1\times10^1$
Pseudomonas C[c]+	2.5×10^9	0.4×10^7
Soil bacteria[c]+	5×10^7	4.2×10^7
S. typhimurium	$<1\times10^3$	$<1\times10^1$

[a] The contaminating bacteria were added to steady state cultures of *Pseudomonas* C and continuous culture was maintained for a further 100 to 200 h

[b] These bacteria were grown together for 1150 h at a dilution rate of $0.1\ h^{-1}$, and then 1.5×10^7 *S. aureus* ml^{-1} were added and the dilution rate was raised to $0.3\ h^{-1}$. Bacteria were counted after an additional 144 h

[c] These bacteria were grown together for 800 h at a dilution rate of $0.1\ h^{-1}$, and then 3.5×10^7 *S. typhimurium* ml^{-1} were added and the dilution rate was raised to $0.3\ h^{-1}$. Bacteria were counted after an additional 168 h. (Rokem et al. [600] with permission)

In experiments directed to test this problem, pathogenic bacteria, *Staphylococcus aureus* and *Salmonella typhimurium*, were inoculated separately into a fermentor containing a monoculture of *Pseudomonas* C growing at a steady state on methanol [600]. The results (Table 22) indicate that *S. aureus* (less than 0.1% of the total number of cells in culture), and *S. typhimurium* (7% of the total number of cells) remained in the culture broth in the growth vessel independent of the dilution rate and without affecting the concentration of *Pseudomonas* C. Furthermore, the concentration of the pathogenic bacteria on the walls of the culture vessel was 5 to 8 times higher than the concentration of *Pseudomonas* C, whereas the ratios in the broth were very different (Table 22). It can be speculated that at least for *S. aureus*, its concentration on the walls might be enough to support most of the bacteria cells found in the culture liquid. In a large fermentor where the ratio of wall area to volume is much smaller than the laboratory-scale fermentor used in this experiment, thousands of bacteria/ml might still be present in the broth.

When substantial inocula of these pathogenic bacteria were added to continuous mixed cultures composed of *Pseudomonas* C and isolated soil bacteria, the pathogens were apparently washed out and were not able to establish themselves either in the broth or on the fermentor walls (Table 22). This suggests that the growth niche of the pathogenic bacteria had been filled by the other heterotrophic bacteria, making the mixed culture less susceptible to invasion by pathogens.

Harrison [607] has stated that although most commercial processes are likely to continue to be based on monocultures, SCP may well be produced using mixed cultures especially from complex substrates. The use of lignocellulosic wastes as fermentation substrates may stimulate the development of mixed culture systems providing a less controversial alternative to combinants of several species formed by genetic engineering techniques. We can expect to see further developments in this area, especially of constructed mixed cultures that are well defined, and therefore can be patented, and are in accordance with IUPAC guidelines [578].

2.2.7 Recycling

Until recently, the recycle of solid (microbial), liquid or gaseous phases in microbiological processes has seldom been practiced. One notable exception being the activated sludge process for wastewater treatment where recycling of a large fraction of the microbial phase is essential for process stability and performance. During recent years, economic considerations were focused mainly on raw materials, nutrients and energy costs, as well as on pollution control. These factors led to the application of recycling technology, especially in commercial plants for SCP production. Today the operation of a large-scale SCP production process involves recycling and reuse of gases, process water/medium and cell biomass. The present diversity of approaches to recycling in fermentation processes have been described in detail by Hamer [614].

Several examples of recycling process water/medium and cell biomass were mentioned in this monograph. The commercial-scale SCP process operated by ICI, incorporates a medium recycling scheme by flocculation and flotation of the

SCP product under aseptic conditions, and also recycling the clear supernate back to the fermentor with added nutrients [569]. The transformation of this recycling concept into an operable plant was most essential for the process, because of the enormous amounts of process water required (from 2 to 2.8 million ton/year; see Hamer [615]), wastage of valuable mineral nutrients and organic byproducts, and the requirement for cleaning large volumes of dilute wastewater prior to discharge.

At Amber Laboratories' commercial plant, protein and/or ethanol is produced from whey in a closed-loop system where no effluents are produced. In this recycling scheme, the concentrated whey is diluted with condensate water derived from evaporation of the fermentation broth, preceding fermentation of the whey. Since the entire fermentation broth is spray dried, no waste streams are obtained [80].

An efficient continuous system for recycling of cell biomass on a laboratory scale was developed at MIT using *Candida utilis* cells grown on spent sulfite liquor [154]. This system improved the productivity obtained without cell recycle dramatically. Furthermore, this work demonstrated the efficiency of on-line monitor and control system for the continuous external recycle-fermentation process.

2.2.8 Recovery of the SCP Product

The recovery process of the SCP product brings the cells from the dilute state in the fermentor to the final state of a dry and stable powder. This process accounts for a significant percentage of the production cost. In terms of the final dry product, quality aspects play an important part in the choice of the recovery process used. The major requirements of the final product include: dry soluble powder, no color or odor, low viable microbial count, no pathogens, high biological values, low RNA or toxic factors and good functionality [616]. These requirements are necessary if the SCP product is to be used as a human food, and especially as a source of protein.

In order to produce an acceptable dry product, the following factors must be considered: low initial solid content from the fermentor, solubility of carbon substrate in the cells (especially in the case of n-paraffins), toxic components in the cells, cell wall digestibility, cell size, cell density, liquid viscosity, removal of flavor and odor of unspent substrates, reduction of viable microbial counts, and type of substrate used [616, 617]. Based on these considerations, numerous schemes for downstream processes were suggested and implemented in industry, and were summarized [616–620]. The schemes include recovery of biomass from the fermentor, concentration (dewatering stage), and drying (dehydration stage), as well as other processes designed to fractionate cell material (cell treatment stage) (Table 23).

Due to the numerous possibilities of recovery methods (Table 23), recovery still remains a major problem in the production of SCP. Most of these methods have not been thoroughly investigated from an economic standpoint, nor from the point of view of what is to be recovered. Some of the suggested techniques are only applicable for laboratory procedures, and not for industrial processing,

Table 23. Recovery and drying methods for SCP manufactures

I. Dewatering		II. Cell treatment		III. Dehydration
Centrifugation		High pressure homogenizer		Sun drying
Filtration – plate, frame, sand, vacuum filter and filter aids		High pressure press		Fluidized-bed drying
Evaporation		Shear in evaporation		Drum drying – single drum double drum
Flotation – foam flotation		Shear in drying		
Sedimentation	Cells →	Freeze thawing	Protein →	Spray drying
Flocculation – autoflocculation genetic modification, pH chemicals: nickel, powder, iron sand, aluminum sulfate, lime, organic cations	Cells → (to III)	Sonication		Tray drying
		Autolysis – chemical, enzymatic		Freeze drying
		Chemical extraction		Pneumatic ring drying
		Ball mill dry cells		

(Based on Labuza [616], Dunnill and Lilly [615], Wisemann [621], and Mitsuda et al. [622])

and in only a few cases were the different methods compared with respect to their cost and the product they yield.

The procedures for SCP recovery that have been incorporated into industrial practice are summarized in the following section.

2.2.9 Description of Selected Pilot Plants and Industrial Processes for SCP Manufacture

Many processes for SCP production employing diverse strains of microorganisms, metabolizable substrates and fermentation equipment are in existence, and have been or soon be operating on an industrial scale (Table 3–6).

In this section, six different processes for SCP production are discussed in detail and compared (Tables 24 and 25). These processes were chosen mainly to demonstrate the use of different microorganisms, the use of different carbon sources as raw materials, and the intended uses of the SCP product. The carbon substrates are: methanol (growing bacteria), sulfite waste liquor (growing molds), CO_2 (growing algae), n-paraffins (growing yeasts), and cellulosic wastes (growing fungi). The SCP product obtained by these processes is to be used for animal feed supplement in contrast to that obtained from glucose (growing fungi) which is used as human food (Table 24).

2.2.9.1. ICI's "Pruteen" Process (Methanol-Bacteria)

An example of an operating commercial-scale SCP process based on methanol is that developed by Imperial Chemical Industries (ICI) Ltd., Agricultural Division

Table 24. Characteristics of selected single cell protein processes

Name of SCP product Item	"Pruteen" bacterial protein	"Pekilo" mold protein	Algal protein	"Mycoprotein" fungal protein	Yeast protein	"Waterloo" fungal protein
Organism	*Methylophilus methylotrophus*	*Paecilomyces variotii*	*Microactinium, Euglena, Chlorella, Scenedesmus*	*Fusarium graminearum*	*Endomycopsis (Candida) lipolytica* FH-H-5027	*Chaetomium cellulolyticum*
Raw material	Methanol	Sulfite liquor	CO_2, domestic sewage	Glucose	n-paraffins	Cellulosis wastes
Company or institute	ICI	United Paper Mills	Technion	RHM	Hoechst-Uhde Gelsenberg	University of Waterloo
Location	Billingham, England	Jamsankoski, Finland	Haifa, Israel	High Wycombe England	Frankfurt, West Germany	Waterloo, Ontario, Canada
Type of process	Continuous	Continuous	Continuous	Continuous	Continuous	Continuous
Fermentor size	1500-m^3	2×360-m^3	Pond area: 120 m^2, 150 m^2, 2×1000 m^2	2×1.3-m^3	4-m^2	1-m^3
Type of fermentor	Air lift	Impeller-agitated	Aerated pond (HRAP)	Impeller-agitated	Air lift	Agitated
Sterility	Aseptic	Aseptic	Nonaseptic	Aseptic	Aseptic	Aseptic
pH	6.5–6.9	4.5–4.7	–	6.0	3.5	5.5
Temperature (°C)	34–37	38–39	Ambient	30	30	37
Specific growth rate (μ) or dilution rate (D) in h^{-1}	0.16–0.19	0.2	2.5–4[a]	0.18	0.11	0.24
Concentration of cells (g/liter; dry wt)	30	17	0.1–0.2	15–20	15–20	–
Y_s-g cell dry wt/g substrate	0.5	0.47	–	0.48–0.5	0.95	–
Productivity (g cell dry wt/liter · h)	4.8–5.7	2.7–2.8	7 kg/m^2 · year	3–4	2	–

Total production (ton/year)	50,000–70,000	10,000	14	50–1000	20	–
Product recovery	Agglomeration and centrifugation	Filtration	Coagulation, flotation, centrifugation	Filtration	Nozzle-type separator, decantor	Filtration
Crude protein (% of dry wt)	83–85	57–63	57	56	58	45
True protein (% of dry wt)	59–64	–	–	50	55	–
Intended for use as:	Animal feed	Animal feed	Animal feed	Human food	Animal feed	Animal feed
Other products formed	CO_2	Water of low BOD	Water of low BOD	–	–	CH_4, CO_2 and water of low BOD
References No.	[7, 8, 55]	[380]	[623]	[8, 45, 624]	[65, 66]	[625]

[a] Retention time, days

Table 25. Comparison between several SCP substrates

Material	Availability		Technical pretreatment	Manufacturing cost (relative units)[a]	Gross energy required (MJ/t SCP)	Competitive substrate uses
	Place	Time				
Methanol	Widespread	Year-round	None	4	117,000	Fuel
Sulfite liquor	Site specific	Year-round	Simple	2	17,000	Decreasing availability
Glucose	Widespread	Year-round	None	6	48,000[b]	Food/chemicals
n-Paraffins	Widespread	Year-round	Separation	5	117,000	Fuel/chemicals
CO_2	Widespread	Variable	None	1	Low	None
Cellulose waste	Site specific	Seasonal	Hydrolysis	3	44,000	Fuel/animal feed

[a] Manufacturing cost increases from 1 to 6.
[b] Data are for molasses (Based on Gaden [364] and Lewis [626])

at Billingham, U.K., for growing *Methylophilus methylotrophus*. This bacterium was isolated from soil and chosen because it is safe, grows rapidly and efficiently on methanol as its only carbon and energy source, and produces a product not only rich in protein but also in the essential amino acids, lysine and methionine [627].

In addition to choosing bacterial strain and raw material, ICI introduced two new concepts in their "Pruteen" process [8]. The first was a radically new design of a fermentor, the "pressure-cycle fermentor". The second was that there would be wherever possible, only one major piece of process plant per process stage. This principle, dictated by least-cost economics, implied that as the process is scaled-up, the fermentor vessel increased in size proportionately. These concepts were examined as of 1973 on a 1000 ton/year pilot plant, and the success of this stage led to the construction in 1979 of a 50,000 to 70,000 ton/year production plant at Billingham (Fig. 24), after spending about $ 100 Million in 11 years of process development [8, 570, 628].

In the general scheme of ICI's "Pruteen" process (Fig. 25), solutions of methanol, ammonia, minor inorganic nutrients, together with water, were sterilized and pumped continuously ($D = 0.16$–$0.19\ h^{-1}$) to the fermentor. The concentration of methanol in the fermentor was approximately 2 ppm, while a suspension

Fig. 24. ICI's "Pruteen" plant at Billingham – a general view. (With permission of John Brown Engineers and Constructors Limited)

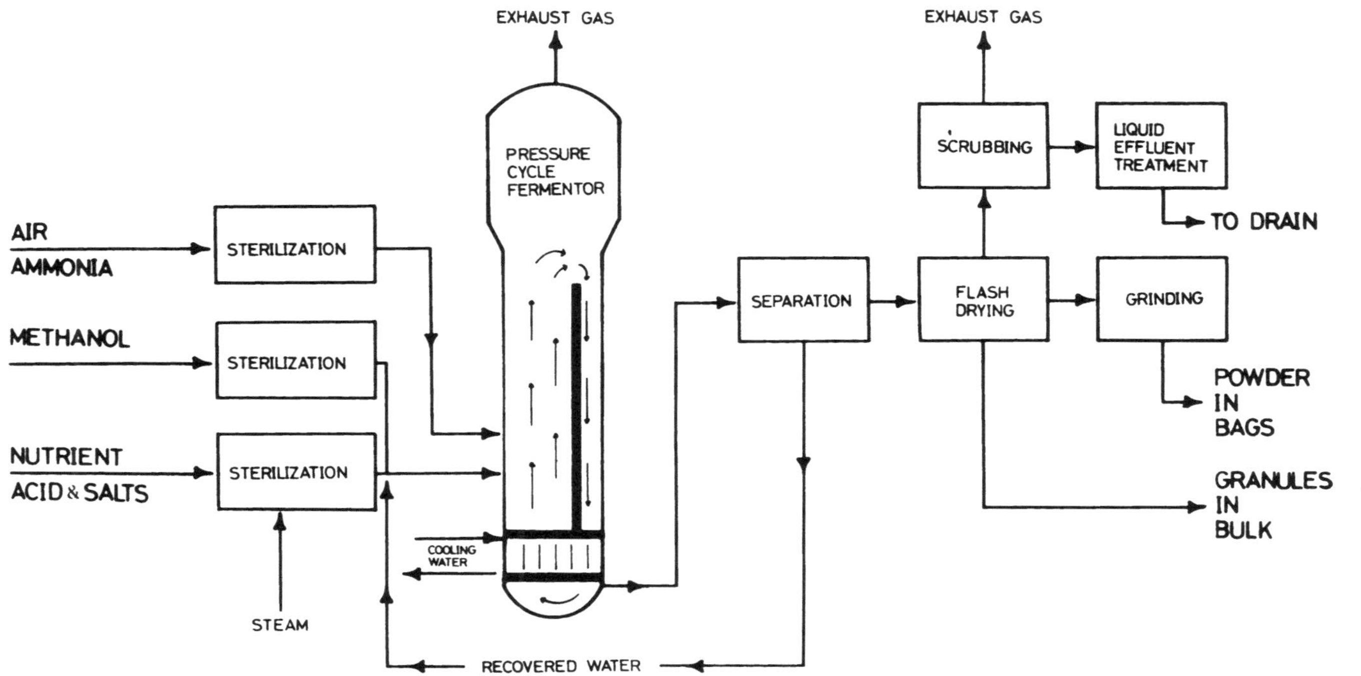

Fig. 25. Main steps in the "Pruteen" process. (With permission of John Brown Engineers and Constructors Limited)

of cells (3 percent dry weight) was continuously removed from the fermentor in the form of a thin slurry. Excess air and carbon dioxide, produced during the fermentation process, were vented from the fermentor through a sterile trap which prevented contamination from outside air; the CO_2 thus obtained is collected and sold as a byproduct.

The cells were separated from the liquid by flocculation and flotation (effected by treatment with chemicals and heat), and were obtained in the form of a thick foam. The liquid was returned, via the sterilizer, back to the fermentor. The thick foam was further dewatered by centrifuges to produce a cream of 25 to 30% cell dry weight, and then passed through a pneumatic hot air drier, resulting in a granular or powdered cell mass ("Pruteen" product).

The "Pruteen" air lift fermentor is the largest aerobic, aseptic fermentor in the world (Fig. 26). It has a capacity of approximately 1500-m^3, uses a 42 m high column (L:D ratio of 6:1) with two chord downcomers, and is driven by 80,000 Nm^3/h of air fed to the base of the riser, with an over-pressure of several atmospheres [461]. The rise of expanding bubbles against a hydraulic gradient is the sole source of power input to the fermentor. The voidage in the riser is greater than that in the downcomer, and the effective density difference provides the driving force for liquid circulation. The riser section is baffled at regular intervals to promote break up of the bubbles.

A large industrial reactor cannot provide a perfectly homogeneous environment, and the problem of uneven distribution of methanol, which is toxic to the organism at high concentrations, was encountered at the pilot plant stage. This problem necessitated the design and installation of a complex nozzle system which introduces methanol through 3000 outlets in the commercial fermentor [569]. The tall air lift fermentor has a major inhomogeneity in that the oxygen supply rate to the culture is reduced by a factor of about 2 on ascending the liquid in the vessel. As oxygen is used, the oxygen partial pressure falls at a rate greater than that to be compensated by the increase in voidage and interfacial area [461]. In order to overcome this problem, the organism used has to be able to cope with the varying oxygen supply rate.

Two other major problems which ICI faced in satisfactorily operating the plant were how to achieve continuous sterility and how to avoid foaming [570]. The first problem was partially solved on February, 1980, by the use of two large computers to monitor and control the continuous sterilization processes [629] and the solution for the second problem was probably a modified dimethyl siloxane antifoam, developed by ICI Organics Division for the brewing industry [570].

Dried "Pruteen" contains 73.8% crude protein (64% true protein) and 3600 kcal/kg metabolizable energy, making it a highly concentrated protein/energy feed ingredient. ICI saw the main markets for "Pruteen" in feeding chickens, pigs, and veal calves; the granular formulation of the SCP product would form an aqueous suspension suitable as a milk substitute [570].

At the end of 1983, granular "Pruteen" was selling at more than \$ 600/ton, twice the European soy price. Clearly with such low protein prices, "Pruteen" can not compete with soy protein. Therefore, ICI is trying both to sell the knowhow to other countries (USSR, Mexico and oil-rich countries), and to explore the use of "Pruteen" as a human food [570].

Fig. 26. ICI's 1.5 million liter fermentor for "Pruteen" production – lifting fermentor into position. (With permission of John Brown Engineers and Constructors Limited)

2.2.9.2 The "Pekilo" Process (SSL-mold)

The "Pekilo" process, developed at the Finnish Pulp and Paper Institute [77, 380], uses spent sulfite (SSL) liquor and is the first commercial continuously operating process, in which the product, the filamentous microfungus *Paecilomyces variotii*, is used as a feed ingredient (Fig. 27). Research work on this process begain in 1963, and it was soon realized that due to the size of this project, extensive cooperation between several companies would be required. A development group, SIT, composed of a number of cooperating Finnish companies, was established for studying possibilities of producing profitable products from SSL. The labora-

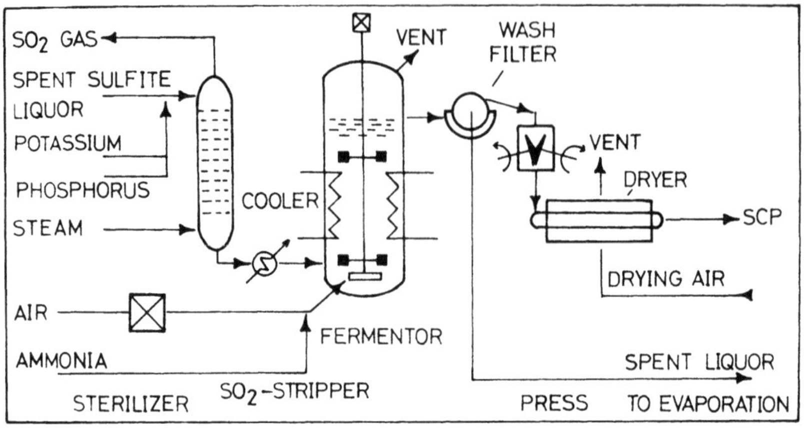

Fig. 27. Simple flow sheet of the "Pekilo" process (Redrawn from Romantschuk and Lehtomaki [77]). (Reprinted with permission of Process Biochemistry)

tory work was followed by a study in a pilot plant (a 15-m^3 fermentor) which was installed in 1973 at the Jamsankoski pulp mill. The results of this pilot plant study were encouraging, and a commercial plant was built and operated in 1974 at the same location.

The Jamsankoski-Pekilo mill has a production capacity of 100 m^3/h of SSL and therefore about 15 thousand ton/year of Pekilo-SCP product can be obtained. The fermentation unit was installed between the cooking department and the evaporation plant of the pulp mill so that while the mycelia resulting from the fermentation are filtered, washed and dried, the spent solution can be returned to the pulp mill's evaporation plant to be burnt. As the microfungi utilize the organic acids in SSL, which are otherwise discharged together with the evaporation condensates, the fermentation process reduces the effluent load and therefore contributes to the protection of the environment.

In the process described (Fig. 27), the SSL is stripped of sulfur dioxide by steam at atmospheric pressure in a tray tower. During the stripping, the pH of the sterilized spent liquor rises, and after cooling, the liquor is fed to the fermentor, together with ammonia, potassium and phosphorus. The process is operated continuously by using two 360-m^3 agitated-aerated fermentors under aseptic conditions; the dilution rate is 0.2 h^{-1}. The mycelial concentration in the fermentor is about 17 g dry wt/liter, with a yield value of 0.55 g cell dry wt/g reducing materials, and a productivity of 2.7 to 2.8 g fungal solids per liter per h. The mycelial product has the advantage over bacterial and yeast cells in that it can be recovered on drum filters in a process that is less costly than centrifugation. The dried SCP product contains 55 to 60% crude protein, of which 87% is digestible; it also contains vitamins and minerals and is used as an animal feed supplement. The suitability of the fungal SCP product for human consumption was also considered after it had been shown [630] that a reduction of the nucleic acid content of the "Pekilo" product was possible (see Chap. 3).

2.2.9.3 High-Rate Algal Ponds (HRAP)

The idea of producing algae for supplementing animal feed, in conjunction with the treatment of sewage in a high-rate algal pond (HRAP) has been studied at the Technion-Sherman Environmental Engineering Research Center, at Haifa, Israel (Table 3) [53, 623]. Israel, having a high level of solar radiation throughout most of the year (latitude 32 ° N), seems eminently suitable for investigating algal growth and processing in outdoor and in field scale pilot plants. This research has concentrated since 1976 on field experiments using ponds fed with domestic sewage. Two ponds – one of 120 m^2 and the other of 150 m^2, are located at the Technion and two ponds of 1000-m^2 each are located near the sewage treatment plant of the Haifa Municipality at Yagur. These high-rate algal ponds are a system of shallow meandering canals equipped with mixing and aeration devices. The ponds are operated continuously throughout the year at optimal conditions: summer-depth 25 cm, retention time 2 days, winter-depth 60 cm, retention time 5.5 days. The symbiotic interaction between algae and bacteria, which is the basis of this process, involves sewage decomposition by bacteria releasing CO_2 while algae, in the presence of solar radiation, use the CO_2 and other nutrients (NH_4^+, PO_4^-), to synthesize cell material. This photosynthetic reaction, besides producing algal biomass, releases oxygen which, in turn, is utilized by the bacteria for sewage decomposition. Such a system, when operated correctly, affords biological sewage treatment without odor nuisances, and provides conditions for maximum photosynthetic activity and thus, algal yield. Some advantages of this dual-purpose process are the following:

- HRAP is a nonaseptic process which does not require fertilizer or additional carbon since municipal waste contains the necessary nutrients for algal cultivation.
- The costs of such a system are offset both by the benefits of waste treatment, the production of biomass, and effluents which can be used for irrigation.
- The flow regime (velocity of about 10 cm/s) and mixing conditions in HRAPs serve to create a continuous mixed culture and increase the efficiency of nutrient utilization for algal production.
- Oxygen is supplemented and controlled by mixing and aeration devices; less oxygen is required for sewage treatment in HRAPs than in aerated lagoons and activated sludge plants, because the main source of oxygen in HRAPs is photosynthesis.
- Sludge treatment is not required as almost no excess sludge is produced; the slight accumulation of sludge is stabilized under anaerobic conditions.

Obviously, the major disadvantages of HRAPs are, the large area required and the limited applicability of the process to places with appropriate climatic conditions.

Based on research carried out for 3 years, several conclusions were drawn:

1) The major components of the suspended matter in the HRAP are algae and bacteria, with minor concentrations of zooplankton and nondegraded matter: The dominant algal species are *Micractinium, Euglena, Chlorella,* and *Scenedesmus.* The ratio of algae to bacteria is generally in the range of 1 to 2.

2) Moderately intense mixing, causing a flow of about 10 cm/s, is high enough to give good algal growth without raising the sediment (short blade rotor mixers proved to be adequate devices for mixing and aeration).

3) The nutrient supply in this HRAP does not limit algal growth during the summer, unless the retention time is too long. A retention time of 2 days seems to be optimal during the summer, and 5.5 days during other seasons.

4) A water depth of 25 cm during summer and up to 60 cm during other seasons, is shallow enough to maintain an adequate algal population in the ponds.

5) The major factor affecting algal productivity is climate, i.e., solar radiation and temperature. The algal concentration fluctuates from 100 mg/liter (dry wt) in winter to 500 mg/liter in summer. The productivity values were 10 ± 2.5 g dry wt/$m^2 \cdot$day during winter and 30 ± 5 g dry wt/$m^2 \cdot$day during summer. The total annual algal production is 7 kg/m^2.

To harvest the algae, solid matter is separated from the liquid by the addition of $Al_2(SO_4)_3 \cdot 18\,H_2O$ (150 ± 50 mg/liter) as flocculant and the algal slurry is floated by dissolved air. The flotator removes suspended solids with between 90 to 94% efficiency. The harvesting process is continued by a settling stage (4 h), resulting in the separation of the algal slurry into two layers, the upper containing a thickened float of 6 to 8% solids, and the lower (drainage) containing 0.2% solids. The algal float with 7% solids is dewatered by centrifugation (solid content of 15%) followed by drying in a double drum dryer. A dried algal moisture content of 10% is obtained.

The final algal SCP product contained 57.4% (of dry wt) crude protein, with protein absorption coefficient of 65 to 80%, and metabolizable energy of 2778 kcal/kg, which almost equals that of fish-meal (2780 kcal/kg). Its amino acid content of arginine, lysine, methionine, histidine, leucine, phenylalanine, and threonine is higher than that found in soybean protein. The SCP product can constitute at least 25% of fish diet and 7 to 10% of poultry diet with no toxic effects.

The effluent obtained from the HRAP, after recovery of algal biomass, can be used directly, or after pretreatment including filtration and chlorination for irrigation.

2.2.9.4 *"Mycoprotein" for Human Food*

In 1964, Ranks Hovis McDougall (RHM) Research Limited, began investigating the production of microbial protein from fungal sources for human food use (Table 5) [8, 45, 624]. Early work was carried out using an isolated soil fungal strain, *Penicillium notatum*-designated C_1, which could utilize starch effluent and hydrolyzed wheat-feed. In 1967, work continued on molasses as substrate, both at High Wycombe (10-liter fermentors) and at Imperial College (400 and 3000-liter fermentors). Initial success led to a pilot plant (1000-liter batch fermentor and 100-liter continuous fermentor) being operated in early 1969, using sugars from potatoes and molasses as carbon sources. At the same time, considerable efforts were made to improve the C_1 strain and to screen for new and better organisms. After screening some 3000 cultures, a *Fusarium graminearum* strain was selected, and proved to be the best in performance both chemically and nutritionally and also possessing a fast growth rate [631].

Although not many details were published, it is known that since the product was intended for human food, the entire process uses only food grade materials in a strictly aseptic continuous fermentation at 30 °C and at a pH of 6.0. Glucose is the carbon source and ammonia (1 g/liter) (the nitrogen source and pH regulator) and phosphate (2 g/liter) are added to the culture medium. The yield obtained is between 0.48 to 0.5 g cell dry wt/g glucose utilized, at a dilution rate of 0.18 h^{-1}, and only a low residual amount of carbohydrate remains in the culture broth. The current scale of operation utilizes two 1.3-m^3 fermentors, producing between 50 to 100 ton/year of SCP product containing 45% protein (of dry weight), of which essential amino acids such as methionine and cysteine comprise up to 3%, with a digestibility value of 93% protein (in laboratory animals).

Recovery of mycelium is considerably less difficult and less costly than that of single cell organisms, and is carried out by filtration after the RNA reduction process. Because of the filamentous structure of the product, it can be used as textured foodstuff without the need for extraction and spinning. It can be used to formulate white meat analogs that are indistinguishable from the natural food, but have substantial nutritional advantages such as low fat, low sodium, high dietary fiber, as well as high protein and low nucleic acid content [8, 632]. This RNA-reduced product (less than 2% of dry weight) known as "Mycoprotein" received U.K. Government clearance in 1980 for sale to the public as human food and is currently undergoing marketing tests [8].

2.2.9.5 Yeast SCP from n-Paraffins

Between 1970 and 1973, a consortium of companies in West Germany consisting of Hoechst, Uhde, and Gelsenberg developed a process for the production of yeast biomass from n-paraffins [65, 66]. After screening about 500 yeasts isolated from soil, water and air, the yeast, *Endomycopsis (Candida) lipolytica* FH-H-5027 was selected from a soil dust sample obtained at a gasoline service station. A mixture of n-paraffins with a chain length between C_{10} to C_{18} was used (at 15 g/liter) as the sole source of carbon. It was found that although this alkane fraction resulted in a better quality product, a fraction of n-paraffins with a chain length of C_{14}–C_{18} resulted in more biomass formation.

The scale-up to continuous culture on a 4000-liter fermentor proceeded via shake flask cultures, and 30-liter and 3000-liter fermentors. The final selection of the desired yeast strain was made after comparison with other strains in terms of the quality of the SCP product and results of experiments with animal feeding. After sterilization, n-paraffins, ammonia, water, phosphoric acid, a potassium hydroxide solution, and antifoam solution are added continuously to the fermentor. Furthermore, magnesium sulfate, zinc sulfate, trace elements, and thiamin are added together with the recycling liquid.

The 4000-liter loop fermentor had liquid jet propulsion with air being supplied through a two-phase nozzle. The liquid jet effected dispersion of the liquid/gas mixture, thus increasing an interfacial area for the O_2 transfer. At the same time, the medium in the fermentor was distributed within the loop by means of internal circulation. There were no rotating parts within the fermentor, thus reducing contamination problems. A continuous fermentation of yeast on n-paraffins (1500 liters) was carried out at a dilution rate of 0.11 h^{-1}, aeration rate of one volume of air per culture volume per min (VVM), at a constant pH of 3.5, and at 30 °C. The cell concentration in the fermentor was between 15 and 20 g/liter, the yield obtained was 0.95 g cell dry wt/g alkane utilized, and the productivity value was 2 g cell dry wt/liter · h. Under optimum operational conditions, the substrate concentration in the spent broth was low (about 0.1%), indicating that the fermentation was carried out under carbon-limitation conditions, which counteracted the accumulation of reserve substances in the yeast cells.

The biomass-containing culture liquid from the production fermentor was fed into a second fermentor which had no rotating parts, and operated as an air lift fermentor. Here, the residual n-alkanes and intracellular storage compounds were further consumed at a greater flow rate (0.2 h^{-1}), and a lower aeration rate (0.3 VVM), thus lowering the residual and unutilized substrate concentration to 0.01% and increasing the protein quality of the product.

A nozzle-type separator was used in the first step of mechanical dewatering and supplied a yeast cream with 15% dry matter. The effluent from the separator was sterilized and recycled into the production fermentor. The 15% yeast cream from the separator was mixed with fresh water in order to reduce the salt concentration of the product. The yeast cream was then subjected to secondary mechanical dewatering in a decanter, which supplied a yeast paste with 25% dry matter and the effluent was then returned to the separator. The yeast paste was heat treated (70 °C for 10 min) to kill the cells and to improve digestibility of the product, then spray-dried to obtain a powder (slighly yellow) containing 3 to 5% moisture and consisting of about 60% crude protein (55% amino acids, 5% nucleic acids). A very low residual n-paraffin content (307 ppm) in the SCP product was obtained by this process, and the product was found to be suitable as a fodder constituent of animal rearing.

2.2.9.6 The "Waterloo" Process-Production of SCP from Waste Biomass

The process developed at the University of Waterloo, Ontario, Canada, is based on the cellulolytic fungus *Chaetomium cellulolyticum* and utilizes agricultural wastes such as crop residues of straw, corn stover and bagasse, animal manure, and forestry residues such as wood sawdust and pulp mill sludges. It thus alleviates the problem of environmental pollution frequently generated by these wastes (Tables 5 and 20) [625].

The process uses a three-stage operation which involves: 1) thermal and/or chemical pretreatment of a cellulosic material, 2) aerobic fermentation of the pretreated material with nutrient supplements, and 3) recovery of the SCP product. The main carbon source for the fermentation is the cellulosic material, while the nutrient supplements are obtained from commerically-available fertilizer blends or animal manure. If manure is used, it is pretreated by anaerobic fermentation yielding methane fuel gas as a byproduct, which supplies operating energy.

The cellulosic raw material is pretreated with hot water or dilute alkali (NaOH or NH_3), depending on the feedstock type, and fed (2% w/v) continuously ($D = 0.24\ h^{-1}$) into a 1000-liter conventional, aerated fermentor operating at pH 5.5 and at 37 °C. No other details are available on the fermentation process except that it can be performed in a tubular fermentor, and in semi-solid systems, using either a mono or a mixed culture of microorganisms.

Recovery of the fungal mycelia is accomplished by filtration, and the dried SCP product contains about 45% (of dry wt) crude protein and appears to be satisfactory as animal feed protein rations in terms of safety, digestibility and nutritive value. The scheme for the "Waterloo" process has several unique characteristics which differ markedly from those described for the other processes. Process costs are minimized by using cheap raw materials (waste residues) and by the versatility of the equipment in handling different raw materials. The process employs low-technology operations, and efficient mass and energy exchanges between processing streams. Economy of the process is improved by replacing the fertilizer with anaerobically predigested animal manure as the source of nitrogen, phosphorus, potassium and various essential minerals. The methane fuel gas concurrently produced is used in the process, and the effluent produced is essentially BOD-free water. Capital and operational expenditures required for the manufacture of the fungal SCP product are relatively small compared to those for other SCP processes, which use expensive feedstocks and high-technology operations. This allows for economic operation of relatively small-size plants in medium-size villages in developing countries.

A comparison between the above mentioned processes for SCP production shows several characteristics:

1) The processes operate as continuous cultures under aseptic conditions (except for algal growth) and at moderate temperatures.
2) Although the biomass yields from carbon source are high, the productivities obtained are relatively low, as compared with values obtained in laboratory experiments. In addition, the dilution rates employed in the different SCP processes (Table 24) are lower than the maximal specific growth rates of the different organisms (i.e., ICI uses dilution rates of 0.16 to 0.19 h^{-1} as shown in Table 24, while the μ_{max} value of *M. methylotrophus* is 0.63 h^{-1}; see [275]). This is because high cell densities effect easier recovery processes, and lower D values reduce the hydraulic flows and minimize the residual substrate concentration.
3) The processes employ simple methods for separation and recovery of biomass, using agglomeration, coagulation and flotation techniques when necessary, or by taking advantage of cell morphology, especially in the case of fungi.
4) It appears from data presented in Table 25 that the most attractive substrates available for SCP production, at least with respect to energy requirements, are the CO_2 and carbohydrates rather than the hydrocarbons and methanol [626].
5) Manufacturing cost figures quoted favor the utilization of CO_2 and agricultural (cellulosic) and industrial (sulfite liquor) wastes as compared with hydrocarbons, methanol and glucose. However, it should be noted that the supply

of SSL is decreasing as paper manufacturers increasingly turn to the Kraft or sulfate process, and though there are great possibilities for cellulosic raw materials, processes based on the latter carbon sources are presently uneconomical for industrial-scale production.

2.2.10 Improvement of the Process for SCP Production

This section summarizes some of the current research and development efforts that may lead to significant improvement in the efficiency and economy of presently operated SCP processes. These are related to: 1) type of fermentor, 2) computer-controlled SCP fermentation processes, 3) genetic manipulation techniques, and 4) the use of thermophilic microorganisms.

2.2.10.1 Type of Fermentor

The design of a fermentor defines its mass and heat transfer capabilities, and hence confines any process which occurs therein, to its operational limits. Since the fermentor is the vessel in which the biochemical conversion of substrates into biomass takes place, the design has great impact on the process conditions [259, 548]. The basic requirements for any fermentor design are that it allows both high productivities and high raw material conversions to be achieved at minimum energy consumption, whilst simultaneously allowing the establishment of an environment conductive to the physiological requirements of the process cultures [614].

Impeller-agitated, sparged, baffled, cylindrical fermentors have dominated the fermentation industry largely due to their flexibility, general utility and suitability for aseptic operation [486, 614]. These fermentors were not individually designed to suit the process for SCP production, neither with respect to optimal performance nor with respect to energy input. They were taken from chemical processes and adapted from pioneering large-scale industrial processes for fumaric acid, antibiotics, ethanol and yeast production by submerged cultures techniques in the mid-1940's. Still, it can be seen that for the "low technology" SCP processes, these fermentors are used. Moreover, in certain existing SCP production plants conventional fermentors had been installed and are still operating (Table 24).

The shift of the fermentation industry towards production of SCP from reduced raw materials of gas-oil, n-paraffins, methane, methanol, and ethanol, stimulated research in these high-technology SCP processes, characterized by the large fermentation volume, continuous mode of operation, high oxygen demand, and low energy input. It then became apparent that conventional fermentors could not answer the problems posed by these demands, and new fermentor designs were suggested [259, 461, 614, 633–635]. The basic type of modern SCP fermentors is the air lift loop fermentor. Equipment for additional mixing (for one phase systems) and dispersion (for multiphase systems) can be added to the basic design. This type of fermentor has significant advantages for the task, as can be seen in Table 26.

Table 26. Comparison of fermentors for SCP production

Conventional stirred vessels	Modern loop reactors
High energy costs	Low energy costs
10 kW/m^3 (approximately)	2 kW/m^3 (approximately)
Less defined fluid flow pattern	Quantified fluid flow pattern
Intrafermentor cooling (jacket)	Extrafermentor cooling (plate exchanger)
Complicated heat removal	Easy removal of the fermentation heat load
Limited scale-up	Less limited scale-up
Indefinite mixing – irregular residence times	More definite mixing – regular residence times
Cell yield on oxygen of 10–20%, hence requiring high aeration rate	Cell yield on oxygen of 40–50%, hence requiring less aeration
Batch and continuous operation	Continuous operation
Low yield and productivity	High yield and productivity
Limited control	Efficient control
Moving parts (easily contaminated)	No moving parts
Expensive capital and running costs	Cheaper to install and run

(Based in part on Prave [636])

The development of new types of fermentors is the main area to which novel technologies have been applied. Many of the engineering problems related to fermentor design were realized only at the pilot to commercial scale-up and were closely related to the biological characteristics of the SCP process. Therefore, the choice of fermentors to be used in production plants must take into consideration the key process parameters of reaction system and reactor, as discussed in Chap. 2 of this monograph.

2.2.10.2 Computer-Controlled SCP Fermentation Processes

Yield maximization is one of the main purposes of control systems for single cell protein production. In order to control this process and other fermentations as well, we need to know: 1) the state of the process within a small time increment, and 2) how the microorganism will respond to environmental conditions, that is, a control model is needed for the fermentation [463]. A set of equations, either algebraic, differential, or partially differential, which describes dynamic behavior of a system is referred to as the model. The model is most valuable in understanding the process, and furnishes an insight into possible control strategies. Furthermore, if the model can predict the dynamic behavior of a process, it is a necessity in optimization and simulation. Different models for various biochemical reactions and reactor design are widely documented in the literature [637–639].

Fermentation control to date has been minimal, involving either set-point controlled variables or following predetermined programs. As process optimization, e. g., yield maximization, requires feedback control, we must know how to change the process on demand and not according to preset plans. Until recently, this was limited to measurement and control of pH, temperature, back pressure, foam, agitation speed, dissolved oxygen, inlet volumetric flow, and nutrient addition rate.

Recently, there has been tremendous development of instrumentation to detect and control environmental factors. This advance has reduced the gap between knowledge about the genetics and metabolic behavior of the cells under various environmental conditions [629, 640–643]. As the number of environmental sensors has increased, the need for better and more efficient methods of data collection and interpretation has increased dramatically. In addition, use of new techniques for the control of fermentation process has spread throughout academia and industry. These developments are creating a fruitful field for computer applications in the present fermentation industry. The recent reduction in computer installation costs brought on by the introduction of the microprocessor, has stimulated the utilization of computers in the biochemical engineering field (see the excellent reviews of Weigand [638], Tannen and Nyiri [644], Armiger and Humphrey [645], and Cooney [646]).

It is worth emphasizing that in biotechnology only a few sensors are capable of on-line measurements of biological variables. Although this deficiency is rapidly overcome, it still hampers effective monitoring and control of complicated fermentation processes. One possible novel approach is the implementation and utilization of computers to combine available on-line sensors for the estimation and calculation of biological quantities otherwise unaccessible. Hence, in these cases computers are utilized to perform high-speed, ultra-complicated calculations. These calculations furnish the data required for controlling the process.

Optimization of biological processes usually deals with optimizing a continuous dynamic controllable process. A controllable process can be made to behave in different ways depending on the quantities controlling the system. While regulation is concerned with maintaining individual process loops at preset values, sequence involves stepping through predetermined program events. Optimization is responsible for determining the "best" overall operating conditions as a function of time [645].

There are two basic approaches to using computers in process regulation: direct digital set-point control (DSC) and direct digital control (DDC). With DSC, conventional analog control loops are used to maintain environmental parameters at previously defined values (set points). In DDC, the computer system interacts directly with the manipulated variables of the process. The advantage of DSC is that a computer system failure does not result in total loss of control of the process. However, the modes of control are limited. In comparison, DDC permits greater flexibility and more precise representation, since control responses can be modelled as algebraic functions. It has the disadvantage of requiring backup controllers to cover for computer system failure [629, 645].

In ICI's SCP process, the computer system (DSC) serves three major functions [629]. The first is to monitor many of the process parameters, i.e., temperature, pressures, flow rates, etc., to provide data storage, data output and general computation facilities associated with carbon and gas balances. The second, and most crucial function is sequence control which includes plant sterilization, inoculation, the transfer from batch to continuous operation and the subsequent control of all process operations. Its third function is to indicate cases of deviations from the process set-points or equipment failure.

A fully automated pilot-plant fermentor has been used and tested for the applicability of DDC technology to the control of a fermentation process [647]. A mixed yeast-lactic acid bacteria population on starch has been used to exemplify the various advantages of DDC coupling.

Several works have been published on optimization and on-line control of SCP production. They described the indirect measurements of cell mass concentration, cell growth rate and substrate concentration [645]. Some of the methods described have been derived by performing material balances around the fermentor via two approaches. One approach selects a specific molecular component and uses a kinetic model relating cell growth with material balance. Zabriskie [648] estimated biomass concentration by a material balance for oxygen corrected for carbon dioxide evaluation, incorporating the function of the respiratory quotient. This author also estimated the biomass produced by using CO_2 in an analogous procedure to the oxygen balancing. A second approach which involves material balancing is based upon the principle of mass conservation [649]. The method is based on the assumption that the elemental compositions of the cells and extracellular products are known. The fundamental relationship states that biomass and extracellular products are produced in stoichiometric ratios. The substrate, along with oxygen and ammonia, is converted to biomass, product, carbon dioxide and water. Once the elemental composition of the biomass and the product have been determined, it is possible to utilize the stoichiometric relationships along with monitoring the oxygen uptake rate, and the substrate or ammonia consumption rate, to continuously monitor the biomass concentration and growth rate.

The method for determining biomass by oxygen uptake [648] has been combined with Wang's technique for material balancing [649] to derive a method of estimating the substrate (i. e., cellulose) concentration in a fermentation process that otherwise is impossible to measure on-line [645, 650]. This method was used for direct digital control (DDC) of glucose medium addition in the production of baker's yeast [651, 652].

The few examples cited above, though representing but a fraction of the studies in this area, illustrate the potential of an on-line computer system for the indirect estimation of important fermentation process parameters. Still, the greatest challenges for future application of computers to fermentation are involved with interpreting results rather than physically implementing control decisions. This must wait until improved techniques for measuring important process parameters on-line and improved mathematical models and strategies for process control and optimization are developed.

2.2.10.3 Genetic Manipulation Techniques

The application of genetic approaches to SCP production has great potential for improving the properties and lowering the cost of single cell protein for use in food or feed. The current research and development efforts may lead to significant improvements in several aspects of SCP production including:

- Hydrolysis of sugar polymers and utilization of the resultant sugar monomers by one organism

- Utilization of all constituents of a multiple carbon source by one organism
- Saving on energy required for SCP production
- Increasing the growth rate of a desired organism
- Isolation of mutants able to grow at elevated temperatures
- Changing the morphological properties of the cells to improve recovery
- Alterations in the amino acid composition of the SCP product.

These are discussed further below.

- Improvement in the economy of SCP fermentation based on sugar polymers (starch and cellulose) is possible by obtaining a microorganism with saccharolytic and/or cellulolytic capabilities, together with the ability to utilize the sugar monomers (both hexoses and pentoses). With the aim of enabling bacterial or yeast strains to convert starch or cellulose directly into SCP in a single stage process, it is desirable that genes for α-amylase, glucoamylase or cellulase be cloned into these organisms [653, 654]. In addition, as various yeast strains are not able to grow on pentose sugars as a sole carbon source, the genes for pentose uptake and utilization should be cloned into these yeasts (see the recent works of Ho and his colleagues [655] and Lawlis and his colleagues [656] on cloning and sequencing of the xylose isomerase and xylulose kinase genes of *Escherichia coli* in this bacterium and in the yeast *Saccharomyces cerevisiae)*. The advantages of this approach are: minimal requirements for pretreatment of the carbon source, and combined sugar polymer hydrolysis and simultaneous conversion of both pentose and hexose sugars to biomass [41]. This, being a new concept, requires evaluation of performance to prove its value.

Cellulase is a repressible enzyme system, and the accumulation of glucose during cellulolysis would attenuate the production of the enzyme and, therefore, limit the growth of both bacteria and fungi. The problem of cellulase repression was circumvented by the use of a cellulase derepressed mutant of the soil microorganism *Cellulomonas* sp.; able to grow efficiently on crystalline cellulose [657, 658]. Also, a mutant strain (D1–6) of *Trichoderma reesei* QM 9414 was found to be partially resistant to catabolite repression and was a hyper cellulase producer [659], similar to the thermophilic actinomycete, *Thermomonospora curvata* [660]. Comparing *T. reesei* mutants from various laboratories, D1–6 strain was approximately 27% better in reported productivity of cellulase. Similarly, the production of α-amylase by yeasts and other microorganisms may be subject to carbon catabolite repression. The use of a non-metabolizable analog of glucose, 2-deoxyglucose, permitted the selective isolation of a relatively large number of α-amylase derepressed and hyperproductive mutant strains of *Lipomyces kononenkoae* IGC 4052 [661].

- Utilization of several constituents of a multiple carbon source by one organism has been demonstrated by Chakrabarty's "superbug", designed to degrade oil [595, 662]. Amanda M. Chakrabarty, while working at the General Electric Company, had combined a number of characteristics from different naturally occurring *Pseudomonas* species into one new strain that could metabolize the major hydrocarbons of petroleum, i. e., camphor (CAM), octane (OCT), xylene (XYL) or naphthalene (NAH) [663]. This case proved to be one of the landmarks in genetic engineering history due to the U.S. Supreme Court's 1980 decision to permit, for the first time, patenting of an artificially constructed strain of *Pseudomonas* that did not exist in nature.

Similar work was described by Portis et al. [664]. Mutants capable of utilizing octane as the sole carbon source were isolated from *Pseudomonas aeruginosa* BS313. Plasmids for the biodegradation of camphor and naphthalene were transferred by conjunction to a *P. aeruginosa* BS316 mutant. The resulting strain, *P. aeruginosa* BS315 was capable of degrading aliphatic, aromatic, and cyclic oil hydrocarbons.

– An interesting application of recombinant-DNA techniques in improving the energetics of ammonia-nitrogen metabolism and, thus, increasing the yield of SCP obtained was developed by ICI's scientists [569, 665, 666]. With the bacterium *Methylophilus methylotrophus* AS1, ICI achieved the high yield of cells growing on methanol as a sole source for carbon and energy, required for the economic production of "Pruteen". After a long period of research, the scientists realized that improvements in efficiency were unlikely to be achieved by manipulation of the fermentation environment alone, and so they turned to the techniques of genetic engineering. Biochemical studies at ICI showed that this methylotroph used an energy-consuming pathway for ammonia assimilation into proteins, consisting of the enzymes glutamate synthase (GOGAT) and glutamine synthase (GS) catalyzing the following reactions:

$$\alpha\text{-ketoglutarate} + NADPH_2 + \text{glutamine} \rightarrow 2\ \text{glutamate} + NADP^+, \quad (12)$$

$$\text{glutamate} + NH_3 + ATP \rightarrow \text{glutamine} + ADP + Pi. \quad (13)$$

Other bacteria are known to possess an energy-conserving nitrogen pathway involving glutamate dehydrogenase (GDH) which catalyses the reaction:

$$\alpha\text{-ketoglutarate} + NADPH_2 + NH_3 \rightarrow \text{glutamate} + NADP^+. \quad (14)$$

A decision was made to conserve some energy during growth by replacing the GOGAT function in ASl strain with a foreign GDH gene. A mobil plasmid pTB70, constructed in vitro from another plasmid into which the SAL 1 GDH fragment was cloned at the SAL 1 site, was successfully employed for cloning the *E. coli* GDH gene into *M. methylotrophus* ASl. A temperature-sensitive glutamate synthase mutant of ASl was isolated, which grew at 30 °C but not at 37 °C. The mutant which could not convert ammonia to proteins, was used as the host. The *E. coli* glutamate dehydrogenase gene (GDH) was cloned by introducing fragments of normal *E. coli* chromosomal DNA, spliced into a plasmid vector, into an *E. coli* glutamate auxotroph, which then was able to utilize ammonia. In the final step it was possible to transfer the cloned GDH gene by mating with the ASl auxotroph that lacked GOGAT. Transconjugants were selected which grew without glutamate at the restrictive temperature. The resulting *M. methylotrophus* strain, equipped with GDH instead of GOGAT, showed a 3 to 5% increase in yield, proving that the methanol (carbon)-limited growth of this organism was, in effect, energy-limited. This modest increase in biomass yield, due to energy efficiency is most valuable in improving the economy proportionally when production is carried out on an industrial scale. A word of caution must be added. Although in laboratory experiment this study led to a significant improvement in yield, the genetically constructed organism is not stably maintained under large-scale process conditions. The basis for this instability might be the genetic inhomogeneity of the population and the selection of a specific genotype better adapted to large-scale cultivation. Hence, there is a need for a rigorous analysis of the performance of recombinant organisms under process conditions [667].

Moore et al. [668] have isolated prime plasmids from *Methylophilus methylotrophus* ASl using the IncP-l plasmid pM0172. Such plasmids can complement mutant function when transferred to appropriate strains of *Pseudomonas aeruginosa* PAO. From a range of particular functions complemented by each prime plasmid, a preliminary map of the *M. methylotrophus* ASl genome was obtained. These authors have suggested that mapping by complementation may be applicable to methylotrophic bacteria in which mutants suitable for selecting recombinants, such as auxotrophs, are difficult to isolate [669]. In addition, identification of different functions in *M. methylotrophus* could be of importance in establishing whether any metabolic functions are dormant or repressed [670, 671].

– Most of the commercially used strains of *Saccharomyces cerevisiae* are polyploid or possibly aneuploid, from which it can be conjectured that supernumerary chromosomes, for one reason or another, are advantageous. James and his colleagues have recently succeeded in constructing triploid and aneuploid strains of the galactose fermenting yeast *Pachysolen tannophilus*, showing that increased ploidity correlated with appreciable enhancement in growth rate on this carbon source [672, 673]. The technique for producing polyploids involved prototrophic selection, and interruption of the normal sequence of events, leading from nuclear fusion to meiosis.

It is hoped that this technique can be applied to other yeast strains for enhancing growth rates on other carbon sources.

– A stable variant of *Methylomonas methanolica*, with a higher temperature optimum for growth, was obtained after mutagenic treatment and selection [506]. The mutant strain Ml3V has an optimum growth temperature of 35 to 37 °C and a maximum growth temperature at 43 °C as compared with 30 and 40 °C for the wild strain. Strain Ml3V and the wild type *M. methanolica* have similar basic characteristics and cell composition. An extracellular polysaccharide is produced by both strains, but this property is more pronounced in strain Ml3V; the extracellular polysaccharide acted as a flocculating agent.

– The application of genetic approaches has great potential for improving the recovery process of the cells from the fermentation broth. The cell-division-cycle (CDC) of temperature-sensitive mutants of *Saccharomyces cerevisiae* A364A were investigated as a means of altering the morphological characteristics and subsequent physical properties of single cell protein [674]. After being grown at 30 °C, and then transferred to 37 °C for 8 hours, strain 4471 formed a visible complex mass at the nonpermissive temperature. Microscopic observations showed that the mother cell was unable to complete the budding process at the nonpermissive temperature, which caused the cells to enlarge and thus increased the settling rate by 25%. This study showed that the physical properties of yeast suspension can be altered by temperatures after propagation, which could lead to a reduction in recovery costs, although, it should be noted that the extra costs of heating the suspension from 30 to 37 °C may still make the process unecomomical.

Another means of reducing the recovery costs is by flocculation, a process by which yeast cells rapidly adhere in clumps and sediment from the fermentation medium. In a flocculant haploid yeast strain, a single dominant gene for flocculation has been identified [675–677]. Mapping studies have revealed that it is located on chromosome 1, and has been designated FLO-1. A further dominant FLO gene has been identified that is not allelic with FLO-1 [678]. Differences in the flocculation characteristics of various yeast strains are primarily manifestations of differences in the yeast cell wall structure, i.e., in its outer surface "hairiness". Using the spheroplast or protoplast fusion technique, Russell and Stewart [679] have successfully fused haploids of *S. cerevisiae* and *S. uvarum* (*carlsbergensis*) strains, the products of which were found to contain the genetic characteristics of both parents. A nonflocculant yeast strain was fused with a FLO-1-containing haploid that possessed a number of nutritional requirements for growth. The recombinant was flocculant and had no nutritional deficiencies. A similar result was obtained by mating an haploid MAT, a strain derived from a nonflocculant commercial baker's yeast (*Saccharomyces cerevisiae*), with a *Saccharomyces cerevisiae* 1209α FLO-1 strain, resulting in the selection of a diploid and extremely flocculant strain [680].

As was pointed out by Stewart [678] in his comprehensive review on the genetic manipulation of industrial yeast strains, fusion is a nonspecific method to genetically modify existing industrial yeast strains in a controllable fashion. Nevertheless, it is a promising technique for production of new hybrids for industrial use (see also the review of Spencer and Spencer [681]).

– Due to the potential economic significance of obtaining SCP with an increased level of some amino acids, particularly methionine, cysteine or tryptophan, consideration has been given to various genetic strategies for altering the amino acid pattern of SCP (these studies are discussed in Chapter 3 dealing with the composition of the SCP product).

2.2.10.4 Thermophilic Microorganisms

Heat is produced as a result of aerobic microbial growth, and consequently, in order to maintain a constant temperature in the fermentor, the heat must be removed during SCP production. The problem can be alleviated by using the cooling tower water or by raising the temperature of the fermentation. The latter approach is potentially less expensive but requires the use of thermotolerant mutants or thermophilic organisms that are able to grow at higher temperatures. The operation of a fermentation process at a high temperature also lessens the probability of contamination, enhances cell recovery as a consequence of decreased liquid viscosity, and provides a product lower in nucleic acid.

Two approaches were used to select microorganisms from the environment which utilized methanol at high temperatures. In the first one, the continuous culture technique was applied and soil samples were used to inoculate the enrichment medium. It yielded in one case a thermophilic mixed population of bacteria having a maximal cellular yield of 0.42 (g cell dry wt per g methanol utilized) at 50 to 56 °C [491]. In a second enrichment experiment a thermotolerant yeast strain, *Hansenula polymorpha* DL-l, was isolated and could grow on methanol at temperatures between 37 and 42 °C, with a yield value of 0.36 g cell dry wt per g methanol utilized [524].

The second approach utilized the batch method and was compared to the continuous culture technique [492]. Although the batch method was outwardly troublesome, it was actually more effective than the continuous enrichment culture through which many prominent yeasts were lost. Using a total of 13,803 soil samples from different places, it was found that soils from tropical regions were rich in thermophilic methanol-utilizing yeasts. A yeast strain, *Pichia methanothermo* 104, isolated from the soil of Ishigaki Island, grew on methanol at temperatures between 37 and 40 °C with a yield value of 0.39 g cell dry wt/g methanol consumed [682].

Although the above mentioned yeasts showed good properties for SCP production, the yields obtained were low. Consequently, any further increase in the cell yield would greatly favor the process of SCP production. Four thermotolerant yeasts growing on methanol were isolated from samples of tropical soils and tropical fruits [522]. One of them, *Hansenula polymorpha* CBM 11, could grow at temperatures up to 40 °C, with a broad optimum temperature range of 26 to 32 °C. It had a relatively high growth rate (0.21 h^{-1}), high cell yield (0.41 g cell dry wt/g methanol) and it could grow at a low pH (3.5), allowing non-aseptic conditions.

A search for thermophilic methane-utilizing bacteria was carried out using a variety of samples from different habitats [286]. Fifteen strains had been selected, and were divided into two physiological groups: thermotolerant organisms capable of growth in the temperature range of 37 to 45 °C with an optimum at 42 °C, and thermophilic strains (50 to 62 °C) with an optimum growth temperature at 56 °C. In a similar approach, a new thermophilic methane-utilizing bacterium (strain H-2) was isolated from a gas field [287]. The bacterium grew at temperatures up to 55 °C, and the optimum temperature for growth was 50 °C. When cultivated in a mixed culture with a heterotrophic bacterium (*Bacillus* strain S), H-2 grew better on methane.

Among 120 isolates of cellulolytic fungi which could grow at 45 °C, *Aspergillus* sp. was selected for its high cellulase activity [683]. As it is important to degrade hemicellulose in addition to cellulose, 180 strains of thermophilic fungi were isolated from waste, and five strains, namely, *Aspergillus, Cephalosporium, Humicola, Talaromyces* and *Thermoascus*, contain high activities of cellulase and xylanase, when grown at 50 °C [684, 685], see also [686].

A great deal of work has been done at the General Electric Company and at the University of Pennsylvania on growth of thermophilic actinomycetes on waste cellulose [402, 687]. One organism, for example, is grown at 55 °C and at a pH of 7.5 to 7.8. Seventy to eighty percent of alkali – or ammonia – treated cellulose is utilized in the fermentation. The problem of downstream processing is that 50% of the protein is extracellular [365].

2.3 Economic Considerations in SCP Production

Among the factors affecting the economic feasibility of SCP processes are:

1) capital costs – cost of land, construction of plant and machinery,
2) site suitability – availability of transport systems, raw material and utilities, size and proximity of markets and safety factors (proximity to urban populations, etc.), and
3) manufacturing (operating) costs – raw materials (carbon and energy source, nitrogen source, and mineral nutrients), energy, water supply, waste treatment, labor, and maintenance, as well as depreciation and the desired profit margin. In many economic estimations the cost attributed to site suitability is included in capital costs.

Numerous publications deal with economic analyses of processes for the production of microbial biomass [8, 33, 38, 58, 547, 626, 688]. However, it is difficult to obtain a realistic economic basis for assessing the relative merits of alternative SCP processes and the advantage of industrial protein production over agricultural protein production [38]. This is due to several factors. Firstly, cost estimates given in the literature are usually supplied by scientists working in academia or for consulting firms, not by those employed by the manufacturers of biomass [39]. In most cases, analyses are made on laboratory-scale or pilot-scale and not on industrial scale processes. Secondly, comparison of alternative routes and technologies tends to be unreliable due to the fact that the effect of inflation on the costs quoted is frequently unquantified, and that some costing exercises refer to commercial technology, whilst others refer only to technology in some state of development [38]. Lastly, any review of economics is liable to be out of date rather quickly due to unpredicted changes in feedstock prices, energy costs, changes in plant construction costs and interest charges [8].

Nevertheless, the data of various economic analyses give sufficient detail to enable several general conclusions to be drawn.

Estimates of the investment required for producing feed-grade SCP from methanol were in the range of $ 600 to $ 1000 per ton of SCP at mid-1978 prices for a plant producing 100,000 tons dry product per year [58]. If we take the total capital investment for a methanol-based SCP process as 100 units, then the relative costs of other processes can be established (Table 27a). The principal variations between these processes will be with the fermentor and fermentor ancillaries, and can be assessed in relation to the required mass and heat transfer capacities [38]. It is apparent that the various high-technology SCP processes are capital intensive, so that only big companies, with substantial government assistance, will be able to construct such plants. This is probably not true for low-technology "village-type" processes, which may operate on a much smaller scale and require more labor (Table 27b).

In the case of relative manufacturing costs, differences between the processes will occur in both the feedstock cost and in the cost of utilities (Table 27). A significant production cost is for raw materials (15 to 77%), primarily for the carbon substrate (17 to 64%) (Table 27 and Fig. 28). A particular element of uncertainty

Table 27. Comparative costs of various SCP processes

a) Capital and manufacturing cost (%) for several SCP processes

	Relative cost (units) for:			
	Methanol	Ethanol	Methane	n-Paraffins
Total capital cost	100	97.6	150.8	107.7
Total manufacturing cost	100	229.6	75.2	129.7

(Adapted from Hamer and Harrison [38])

b) Comparative production costs of SCP

Process	Relative production costs (%) for:					
	Raw materials:		Utilities	Labor	Miscel-laneous	Total
	Substrate	Other chemicals				
Yeast-paraffin	29.4	23.9	23.8	8.4	14.5	100
Bacterium-methanol	47.4	26.4	14.2	6.2	5.8	100
Yeast-ethanol	63.9	9.9	12.0	5.1	9.1	100
Fungus-sulfite liquor	17.0	33.7	24.8	11.0	13.5	100
Bacterium-bagasse	25.7	13.5	36.6	8.3	15.9	100
Algae-CO_2	?	16.5	13.8	15.3	54.4	–
Fungus-lignocellulosic wastes	?	14.5	11.3	22.9	51.3	–

(Results calculated from Moo-Young [688[, Berend et al. [623], and Moo-Young et al. [625])

exists for processes which utilize CO_2 or waste materials. In these cases, the cost of the carbon source may be assumed to be zero or, in some cases, even may be assigned a negative value because of the cost incurred in disposing it through municipal sewage systems or through open field disposal. In some cases, the cost of the carbon source may be attractive for the production of SCP, but the carbon source may not be available in sufficient quantity at a central location and, hence, transportation costs may be prohibitive [39]. Furthermore, the carbon source can be used in other fermentation processes for production of chemicals and energy, and thus the SCP fermentation would have to compete for a carbon source, which has a certain cost value, rather than a zero-cost value.

In most cases, the cost of the carbon source is the single largest contributor to total operating expenses; thus, it is obvious that for the process to be more economic, conversion efficiency of substrate carbon to cell carbon (the yield value) must be maximized. The change in investment and manufacturing costs as a function of cell yield on methanol is shown in Fig. 28. Higher yield values give lower costs due to lower oxygen demand and lower heat produced (cooling requirement) per unit of biomass formed, lower costs of biomass separation and lower quantity of waste water.

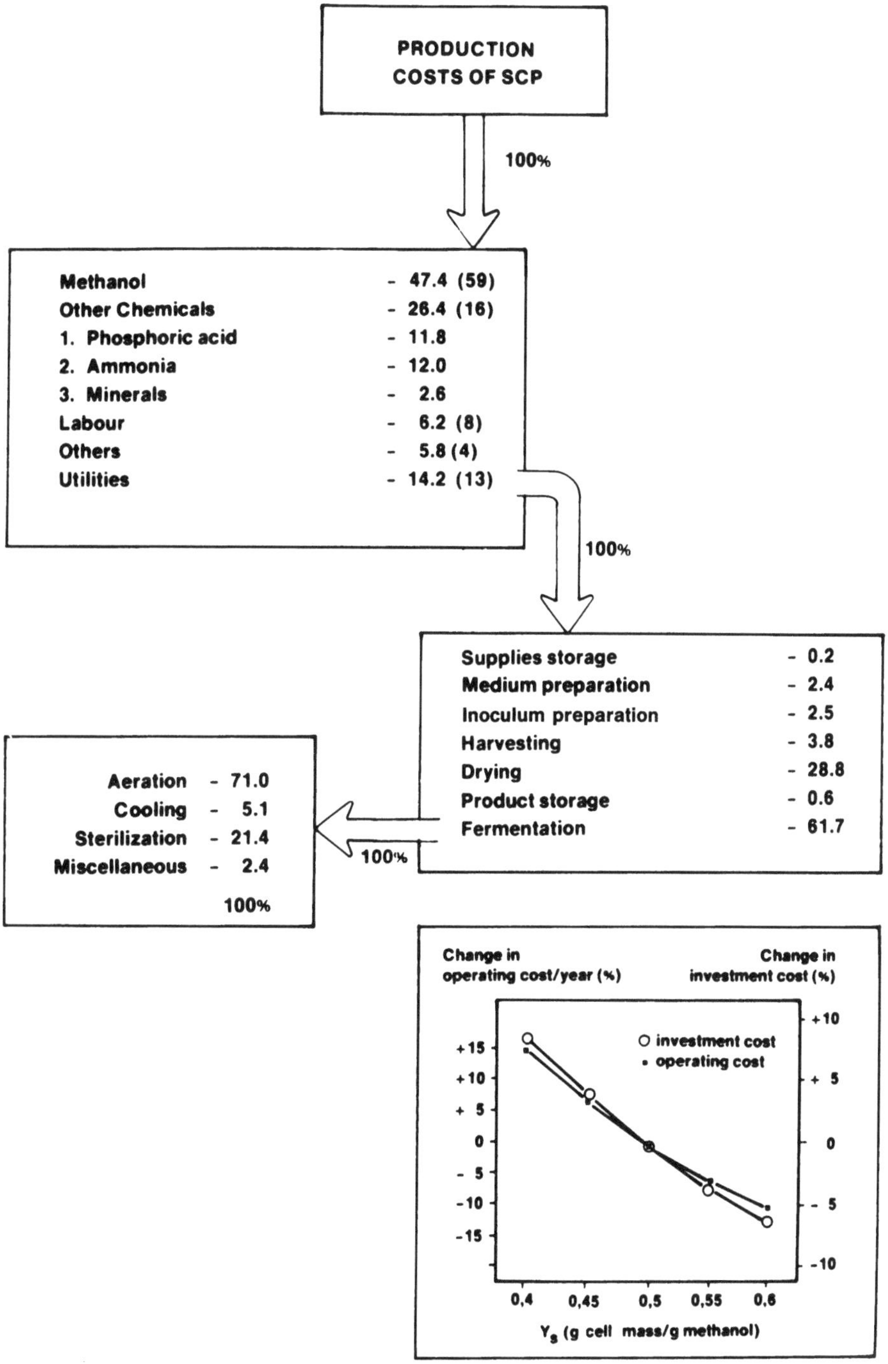

Fig. 28. Distribution of manufacturing costs (%) for the production of bacterial SCP from methanol. Calculated from data presented by Moo-Young [688]. Numbers in parentheses represent results on cost estimate for 100,000 ton/year of the "Norprotein" SCP process [58]. The graph describes investment and manufacturing costs as a function of cell yield on methanol (reprinted with permission from Ebbinghaus et al. [58], copyright © 1981, Pergamon Press)

Of the total utilities cost, fermentation has the highest expenditure, followed by drying and harvesting (Fig. 28). Most of the cost of fermentation facilities lies in the aeration requirement, while cooling demands appear to be relatively small (Fig. 28). High costs for facilities are incurred in sterilization, i.e., in processes that are operated under aseptic conditions.

Estimates of energy requirements for SCP processes vary widely (Table 25). It was concluded that the gross energy requirement of hydrocarbon-based SCP was unfavorable compared to carbohydrate-based processes. The 1973 oil crises caused a marked increase in feedstock prices and energy costs, and thus further limited the relative commercial potential of SCP based on n-paraffins, methanol, methane and ethanol for animal feed supplements [8]. Future prospects for processes based on these carbon sources (in particular methanol and ethanol) but intended as food-grade materials may have a realistic chance of proving to be an economical proposition simply because of the very high meat prices. However, production scales will never approach those envisaged for animal feed purposes, and the SCP product will be limited to use as protein supplements and functional protein ingredients, rather than as a primary source of protein in human diets [7].

Other small-scale, low-technology processes designed for upgrading agricultural residues and other wastes will probably be intended for production of SCP products for animal feed. These processes may find useful applications in those areas where low-cost carbohydrate residues are available, and conventional protein feedstuffs are in short supply [7, 8].

3 The SCP Product

3.1 Introduction

Although the principal nutritional value of the SCP product is contributed by the protein content and the composition of amino acids, particularly essential amino acids, microbial biomass contains other main components such as nucleic acids, carbohydrates, lipids and minerals. The macromolecular composition of the SCP product is basically determined by the genetic make-up of the cell but varies, to a certain extent, as a function of changes in growth conditions. Thus, the ultimate aim should be to select appropriate strains of microorganisms, with optimal growth conditions, for obtaining maximal protein content with a favorable composition of amino acids.

If the single cell protein is to be used as a major part of the product, and not as a minor food or feed additive, e. g., as a source of flavors or vitamins, it should have the following properties: dry soluble in powder form, no color or odor, low viable microbial count, no pathogens, high biological value, low RNA or toxic factors, and good functionality and taste.

In this chapter, the composition of the SCP product and the nutritional and toxicological aspects involved in using this product are discussed.

3.2 Composition of the SCP Product

The principal value of microbial biomass as a component of feeds or foods is its contribution to protein nutrition. However, any SCP product also comprises of peptides, amino acids, amines, nucleic acids, nucleotides, carbohydrates, lipids, vitamins, and minerals (Tables 28–31). Generally, the producer is satisfied if the affects of these non-protein materials are negligible. Vitamins and minerals are routinely added for fortification of foods and feeds, and although they can be obtained rather inexpensively from other sources, they are available here in the SCP as a "bonus". This may also be true for nutritional factors, such as "unidentified growth factors," which significantly improve the growth response of animals to a diet containing complex natural materials such as SCP product [36].

Table 28. Macromolecular composition of selected SCP products (% of dry wt)

Organism	Growth substrate	Crude protein (N × 6.25)	True protein	Crude fat (oil)	Total carbo-hydrate	Total nucleic acid	Ash	Ref.
Methylophilus methylotrophus	Methanol	73.8	64	8.5	3.4	15.9	9.0	[55, 689]
Pseudomonas C	Methanol	75.5	67.0	3.1	7.0	16.0	8.0	[Personal]
Methylomonas methanolica	Methanol	81	67	10.5	–	–	4.8–7.2	[690]
Cellulomonas sp.	Glucose	74	67.6	–	12	28	9.2	[658]
Hydrogenomonas eutropha	Urea, H_2	84	–	–	trace	7.8	5.88	[691]
Candida utilis	Sugar	56	46	1.1	25	13.4	8.2	[692]
Torulopsis candida Y-128	Palm oil	46.1	43.8	1.3	36.5	–	9.8	[132]
Candida guilliermondii	n-Paraffins	63–68	49–54	1.8–3.1	23–27	–	4.8–7.2	[135]
Candida paraffinica	n-Paraffins	58	–	7	–	–	10	[Personal]
Yeast sp.	n-Paraffins	57	48	10	–	11	–	[69]
Candida sp.	n-Paraffins	60.3	–	2.8	4.2	–	11.5	[260]
Candida scottii P-2	Naphtha	45.2	–	3.42	19.3	7.1	6.2	[138]
Monilia sp. R5-H	Kerosene	37–44	–	24–28	25–31	–	3.8–4.3	[211]
Trichosporum pullulans	Gas-oil	52.76	38.5	3.84	22.6	5.23	7.55	[693]
Fusarium graminearum	Glucose	54.06	42.94	0.96	14.68	6–13	6.08	[694]
Paecilomyces variotii	Sulfite liquor	57–63	55	2–4	25	9	5–6	[380]
Chaetomium cellulolyticum	Cellulose	45	–	10	35	5	5	[625]
Scenedesmus	CO_2	50–55	35	8–12	10–15	6.0	8–12	[175]
Spirulina	CO_2	55–65	35	2–6	10–15	3.9	5–12	[175]
Chlorella	CO_2	40–55	35	10–15	10–15	–	5–10	[175]
Soya	–	35–40	–	15–20	20–35	–	4–5	[175]

From Table 28 it is apparent thatSCP compares well with high-quality protein sources such as soya, in terms of the amount of "crude" protein that the product can supply as well as in the overall pattern of nutrients. The protein content is highest in bacteria, which is above 64% true protein. In yeasts it amounts to 44–54%; in filamentous fungi it comprises between 38 to 55%; and in microalgae, the concentration of protein is the lowest – 35%. It appears that among the various types of SCP, products of bacterial origin contain the highest content of nucleic acids and the lowest amount of carbohydrates (Table 28). Microalgae appear to have the highest content of lipids, including the plastid pigments – chlorophylls and carotenoids.

3.2.1 Protein

The protein content of SCP can be measured by several techniques which give different results, are of different convenience, and are relevant to different aims [36]. The term "crude protein", defined as total Kjeldahl nitrogen × 6.25, has often been criticized because it does not differentiate between protein-nitrogen and non-protein-nitrogen (nucleic acids, amines, urea, and ammonium citrate), and invariably gives a misleadingly high estimation of protein content (PAG Guideline No. 12). The Biuret method is a good measure of protein, simple to use, but requires that the SCP product be dissolved in alkali, which sometimes presents serious problems [36]. The term "true protein" is defined as the sum of the anhydro amino acids determined by direct chemical analysis and is accepted (PAG Guideline No. 6) as an accurate estimation of the amount of protein in a given material; this value is always significantly lower than the "crude protein" value, as is evident from data shown in Table 28. Although the amount of protein based on the sum of amino acids is accurate, the analytical methods used for amino acid determination are rather inconvenient. In addition, various microorganisms, e.g., microfungi and algae, produce small amounts of some unusual amino acids, and their correct identification is necessary for accuracy of this test [699–701].

On the basis of an adequate caloric input, the nutritional protein quality is dependent on the relative proportions of its amino acids, essential to a given organism, and is often compared to the amino acid content of whole egg or the slight modification thereof, that is recommended as a reference standard by the Food and Agriculture Organization (FAO) of the United Nations (Table 29). In the case of SCP, the amino acids which tend to be limiting in terms of the nutritional value are the sulfur-containing ones, while lysine, in which wheat is deficient, is present in optimal concentration (Table 29). However, since the animal species tested, e.g., pigs, hens, and chickens differ in their amino acid requirements, e.g., glycine is essential for poultry but not for pigs, it cannot be expected that a particular SCP product is the general optimal one [701].

In comparing published values of amino acid profiles of different SCP products (Table 29), a degree of caution is required. Most of the reports lack statistical information concerning size of representative samples, number of analyses, number of samples, standard deviation and variability between different batches, and also lack essential information with regard to the analytical procedures for

Table 29. Content of essential amino acids of microorganisms considered for use as SCP (selected list)[a]

Protein source	Nitrogen × 6.25	Lysine	Threonine	Cysteine	Methionine	Valine	Isoleucine	Leucine	Phenylalanine	Tryptophan	Growth substrate	Ref.
FAO[b] Reference	–	4.2	2.8	2.0	2.2	4.2	4.2	4.8	2.8	1.0	–	[36]
Soya bean meal	45.5	2.8	1.9	0.7	0.6	2.3	2.2	3.5	2.2	0.6	–	[689]
Fish meal	66.3	4.9	3.0	0.7	1.9	3.7	3.2	5.0	2.9	0.9	–	[689]
Methylophilus methylotrophus (Pruteen)	73.8	4.7	3.4	0.4	1.8	3.9	3.3	5.1	2.6	0.8	Methanol	[689]
Pseudomonas C	75.5	8.3	5.9	1.1	2.6	6.7	5.7	8.1	4.0	–	Methanol	[512]
Methylomonas methanolica (Norprotein)	81.0	4.8	3.6	0.5	1.7	4.7	3.7	5.9	2.9	–	Methanol	[690]
A bacterium sp.	73.4	3.9	2.9	0.1	1.3	3.8	2.7	3.9	2.9	–	Methanol	[695]
Candida lipolytica (Toprina)	66.0	7.8	5.4	0.9	1.6	5.8	5.3	7.8	4.8	1,3	n-Paraffins	[593]
Paecilomyces variotii (Pekilo protein)	57–63	6.4	4.6	1.1	1.5	5.1	4.3	6.9	3.7	1.2	Spent sulfite liquor	[380]
Saccharomyces cerevisiae	53.7	8.2	4.8	1.6	2.5	5.5	5.5	7.9	4.5	1.2	Molasses	[696]
Kluyveromyces fragilis	56.2	6.9	5.8	–	1.9	5.4	4.0	6.1	2.8	1.4	Cheese whey	[696]
Candida utilis (Torutein)	48	7.1	4.7	0.6	1.0	5.3	4.3	7.0	3.7	0.04	Ethanol	[39]
Trichoderma harzianum (ICAITI protein)	–	4.7	3.8	–	1.1	4.9	4.0	5.8	3.5	–	Coffee waste	[456]
Chaetomium cellulolyticum (Waterloo protein)	45	6.8	6.1	2.6	5.8	4.7	7.5	3.8	3.8	–	Lignocellulose	[625]
Monila sp. R5-H	37–44	12.2	7.0	–	1.0	6.2	5.9	8.8	6.8	2.5	Kerosene	[211]
Spirulina	55–65	4.8	6.2	0.9	2.5	7.1	6.7	9.8	5.3	0.3	CO_2	[175]
Hydrogenomonas eutropha	84	8.6	4.5	–	2.7	7.1	4.6	8.5	4.0	–	Urea, H_2	[691]

[a] Data are g amino acid per 100 g protein or 16 gN
[b] FAO – Food and Agriculture Organization of the United Nations

protein hydrolysis and determination of amino acids [36]. However, it can be stated that bacteria generally have a relatively high content of methionine and cysteine and a low content of lysine, the opposite holds for yeasts, and algae have relatively high leucine concentrations.

3.2.2 RNA and DNA

All types of SCP products contain much higher amounts of nucleic acids (up to 16% of dry wt; see Table 28) than "conventional" foods or feeds. Since all animal species, except man and anthropoid apes, metabolize purines beyond the uric acid stage to allantoin, this high nucleic acid content is unfavorable only when the SCP product is considered as human food. This was shown by Endozien et al. [702] and Kihlberg [31] with the finding that dietary nucleic acid is depolymerized by nucleases in the pancreatic juice, and then converted to nucleosides by intestinal enzymes prior to absorption. The final metabolic product formed in man from the purine moiety (guanine and adenine) of nucleic acid is uric acid. The enzyme uricase, which oxidizes uric acid to the soluble and excretable metabolite allantoin, is present in most mammals but not in man. Consumption of a protein source high in purines results in a higher plasma level of uric acid, and an increased excretion of uric acid in the urine. Due to the low solubility of uric acid, it is suspected that if the plasma uric acid content is increased, crystals of uriate may form in tissues and joints, causing gout or gouty arthritis [703].

Possible negative effects in man of consuming pyrimidine bases have also been observed; feeding large amounts of orotic acid resulted in a fatty liver and liver degeneration [703, 704].

3.2.3 Lipids

Total lipids (fats) obtained by the usual solvent extraction method vary between 1 and 15% (of dry wt) including triglycerides, phospholipids and sterols (Table 28). Algae contain the highest content of lipids (including the pigments). In most cases, the nature of the lipid fraction has apparently not been determined. In methanol-grown bacteria [705, 706], profound differences were found in the phospholipid composition between bacteria that utilize the C_1-compound via the RMP pathway and those employing the serine pathway. Thus in RMP-pathway bacteria, about 76% of the phospholipids consisted of phosphatidylethanolamine (PEA), and the ratio between PEA and phosphatidylglycerol (PG) was higher than 4. Phosphatidylcholine (PC) could not be detected in extracts of these bacteria (Table 30). In the other bacteria, the content of PEA was about half of that found in the RMP-pathway bacteria, and the PEA-plus-PG contents were between 51 and 70.4% of the total lipid content. The ratio between the contents of PEA and PG was 1.8 to 1.9, and high levels of PC were found [697]. Variations in the composition of fatty acids of the two groups of methanol – grown bacteria were also noted (Table 31) [270, 344, 697, 698, 706]. In those bacteria that utilize

Table 30. Chemical composition of ICI's "PRUTEEN" product and *Pseudomonas* C

Vitamin analyses (mg/kg)		Ash analyses (% of dry wt)
Ca-d-Pantothenate – 1.58		P_2O_5 – 6.58
Folic acid – 8.74		MgO – 0.53
Nicotinic acid – 58		K_2O – 0.30
Choline – 10		Na_2O – 1.53
Biotin – 3.12		CaO – 0.04
Inositol – 50		Fe_2O_3 – 0.12
A – 0.00	B_6 – 0.93	Total ash 9.00
B_1 – 5.75	B_{12}–23.00	
B_2–74.30	E –28.00	

Trace ions (ppm)		Potential harmful ions (ppm)		Phospholipid composition of *Pseudomonas* C (% of total lipid P)	
Mn	12–36	As	0.01	Phosphatidylethanolamine (PEA)	76.2
Zn	8–30	Hg	0.01	Phosphatidylglycerol (PG)	12.8
Cu	3–7	Sb	0.1	Phosphatidylserine	8.8
Se	< 0.1	Pb	0.05	Diphosphatidylglycerol	2.2
Co	< 10	Cd	0.1	Phosphatidylcholine	0
Cl	< 0.05%	F	10	PEA/PG	6.0

(Based on [689, 697])

Table 31. Fatty acid composition of methanol-utilizing bacteria

Source	Metabolic Pathway	Percentage of fatty acid					Other fatty acids	Ref.
		14:0[a]	16:0	16:1	18:0	18:1		
Butter fat	–	11.1	29.0	4.6	9.2	26.7	19.4	[689]
Corn oil	–	1.4	23.4	2.0	1.1	22.9	49.2	[689]
Soy bean	–	0.1	9.8	0.4	2.4	28.9	58.4	[689]
Pseudomonas C	RMP	3.5	39.0	52.6	1.6	1.2	0.8	[697]
Methylomonas methanolica	RMP	4.0	24.2	49.1	2.0	20.6	–	[697]
Methylophilus methylotrophus	RMP	1.2	39.6	45.2	1.0	4.4	8.6	[689]
Isolate K-81	RMP	0.8	40.4	38.2	0.7	5.8	14.1	[698]
Methylomonas clara	RMP	2.0	41.0	30.0	1.7	2.8	22.5	[259]
Pseudomonas 135	Serine	0.4	6.5	1.7	8.6	80.2	2.6	[697]
Pseudomonas M-27	Serine	1.9	4.0	8.5	2.3	78.5	4.7	[697]
Pseudomonas 1	Serine	2.2	3.5	1.0	2.5	87.8	3.1	[697]
Pseudomonas AM-1	Serine	3.6	7.4	6.5	11.6	68.3	2.5	[697]
Pseudomonas rosea	Serine	1.3	3.0	0.5	3.7	87.1	4.8	[697]
Pseudomonas extorquens	Serine	0	4.3	13.1	2.1	72.7	7.8	[698]

[a] Fatty acid methyl esters are designated by the number of carbon atoms to the left of the colon and number of double bonds to the right

[b] RMP, Ribulose monophosphate pathway

C_1-compounds via the RMP-pathway, between 73.3 and 91.6% of the esterified fatty acids were $C_{16:1}$ [1]. The relative content of $C_{16:0}$ was about the same as that of $C_{16:1}$ in these bacteria. In bacteria that utilize C_1-compounds via the serine pathway, between 68.3 and 87.8% of the esterified fatty acid was $C_{18:1}$ (Table 31).

While microalgae contain relatively high concentrations of unsaturated fatty acids, alkane-grown yeasts contain high content of uneven-numbered, cyclopropane and branched-chain fatty acids. These acids might cause adverse effects in animals [701].

Several reports have shown that certain bacteria [707] and blue-green algae [708, 709], are capable of synthesizing steroids and, by implication, squalene, a precursor for steroids, in a mode similar to eukaryotic organisms. The amounts of steroids (or squalene) found in prokaryotes (0.001 to 0.1 mg/g cell dry wt) are, in most cases, considerably lower than those found in eukaryotic organisms; for example, *Aspergillus nidulans* contains 0.3 mg squalene/g cell dry wt [710, 711]. Two striking exceptions are *Halobacterium cutirubrum* [712–714], in which the amount of squalene is 1 mg/g cell dry wt., and the methane-utilizer *Methylococcus capsulatus* [710, 711], which contains 5.5 mg squalene/g cell dry wt. In a comparative study of methanol-utilizing bacteria, we found that different RMP-pathway bacteria contain squalene in concentrations between 0.1 and 1.16 mg/g cell dry wt [715]. Squalene could not be detected in lipid extracts of methanol-grown bacteria which utilize C_1-compounds via the serine pathway.

As will be discussed later, culture conditions affect lipid content, which is increased when the medium contains a surplus amount of sugar or a limited amount of nitrogen [716]. In bacteria grown under nitrogen limitation, poly-β-hydroxybutyric acid (PBH), a common reserve material, may be accumulated to comprise up to 80% of the dry weight [717]. As PBH is indigestible by vertebrates, it is desirable to utilize PBH-free mutant cells for SCP production [718].

3.2.4 Other Constituents of the SCP Product

Cell walls represent between 0.5 and 10% of the cell dry matter; in eukaryotes the percentage is higher than bacteria and is expressed as the crude fiber fraction of the cell. This fraction usually consists of a variety of macromolecules which are group specific, species specific and even strain specific. The typical structural macromolecule of bacterial cell wall is murein. It consists of chains in which N-acetylglucoseamine, and its derivate, N-acetylmuramic acid, alternate. The chains are interconnected by oligopeptide bridges which contain m-diamino-pimelic acid, and, in the case of Gram positive bacteria, teichoic acid. In the cell walls of blue-green algae and microalgae, polysaccharides dominate over murein. In yeasts, glucan and mannan are the main components of the cell wall, with the remainder consisting of chitin (found in fungi), proteins and lipids [719]. The intact, complex polysaccharide cell walls are generally obstacles to digestibility [701].

---

1 Fatty acids are designated by the number of carbon atoms to the left of the colon and number of double bonds to the right

Carbohydrates vary between 3.4 and 36.5% of the cell dry wt (Table 28), with the highest values being found in yeasts. Glycogen can be found as a reserve material in bacteria, fungi and yeasts, whereas starch is the reserve material in algae. Pigments such as chlorophylls or carotenoids constitute 3 to 5% of the biomass of phototrophic microorganisms, e.g., microalgae. The ash content varies between 4.8 to 12% of dry wt (Table 28) and about 74% of the ash content is composed of phosphorus salts (Table 30). Potentially harmful ions and trace ions are present in relatively small quantities (Table 30).

From the nutritional point of view, yeasts were first used as sources of vitamins, because they represent one of the richest sources, particularly of the vitamin B complex group. In *Torula* yeast the concentration of vitamin B, B_2, and B_6 are 5.3, 4.5, and 33.4 mg/kg, respectively, and in Brewer's yeast the concentrations are 50–360, 36–42, and 25–100 mg/kg, respectively [720]. The concentrations of different vitamins in bacterial SCP product are shown in Table 30.

The fear has been expressed that accumulation of aliphatic hydrocarbons by yeasts grown on n-paraffins, and the carry over of residual n-alkanes and carcinogenic polycyclic aromatic compounds (especially 3,4-benzpyrene), might be a hazard to animal health. However, it is now well established that these compounds are so efficiently extracted from the SCP product that their final level is definitely lower than in most "conventional" feedstuffs and foods [721, 722].

3.3 Changes in the Composition of the SCP Product

3.3.1 Changes in Macromolecular Composition

3.3.1.1 Protein and Nucleic Acids

The macromolecular and micromolecular composition of the single cell protein product is determined not only by the cell's genetic make-up, but also by growth conditions and the environment. Cell composition varies, to a certain extent, as a function of the pH, temperature, ionic strength, and the nutrient composition of the medium. In addition, cell composition varies as a function of the growth rate of the organism, due to variation in metabolic needs. Thus, the continuous culture has been widely exploited to study the effect of growth rate on the macromolecular composition of the cell, because with it the dilution rate can be varied, thereby achieving a variety of growth rates [723].

Generally, the highest protein content results from the use of the carbon source as he limiting nutrient; however, exceptions have been described. Thus, *Candida utilis* cells, grown under glucose limitation were observed to decrease markedly in protein content on increasing dilution rate [724]. Such variations are unusual, compared with reported examples. The percentage change in protein content in this organism appeared to be largely due to the increased synthesis of glycogen at the higher growth rates. If other factors, such as nitrogen, oxygen, phosphate or metals, were the limiting nutrient, there was a tendency for the cell to

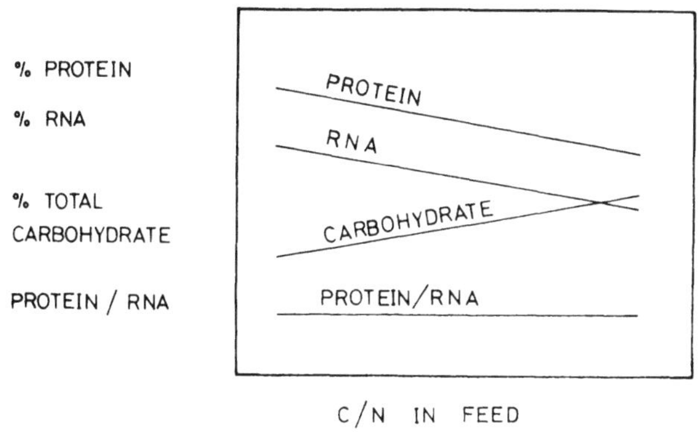

Fig. 29. Cell macromolecular composition, at a constant growth rate, as a function of the carbon/nitrogen (C/N) ratio in the growth medium. (Redrawn from Wang et al. [729]) (Reprinted with permission of John Wiley & Sons, Inc.)

form an excess of other cell components [348, 725–730]. Each of these nutrients may be expected to exert different pressures on the cell physiology when used as the limiting nutrient, since they each play a different role in metabolism. Thus, when carbon (energy) is limited, there will be a restricted supply of carbon for biosynthesis of energy, whereas in nitrogen or sulfur limitation, protein synthesis will be restricted. Limitation of phosphate may restrict nucleic acid synthesis and energy production, and magnesium or potassium limitation may inhibit nucleic acid or cell wall synthesis [463]. It is expected, therefore, that by choosing different growth limiting nutrients some metabolic activities may be restricted while others will continue at the same rate, in effect changing the macromolecular composition of the cell. Cooney et al. [729] observed the pattern shown in Fig. 29, for the macromolecular composition of *Enterobacter aerogenes*, grown under a dual carbon-nitrogen limitation. As the ratio of carbon (glucose) to nitrogen (ammonia) was decreased, the cells shifted from carbon to nitrogen limitation. The protein, RNA, and carbohydrate content of the cells went through a continual transition with sole carbon or nitrogen limitation representing the limits in cell composition. Even though the actual concentration of protein and RNA changed as a function of the carbon to nitrogen ratio, the protein to RNA ratio remained constant. This illustrates the need to consider not only absolute concentrations but also relative concentrations of cell components.

The nature of the limiting nutrient is also important in controlling cell composition. This was illustrated by comparing ammonia and nitrate cultures of *Candida utilis* which were cultivated under otherwise similar conditions [731]. Although the nucleic acid to protein ratio in these cultures was approximately the same, the contents of both protein and nucleic acid were higher in cells fed nitrate, than in cells grown on ammonia. The difference was probably due to a higher content of carbohydrate in the cultures incubated with ammonia. A different result was obtained for the methanol-utilizer *Pseudomonas* C, where a large content of extracellular polysaccharide was produced during growth on nitrate, but not on ammonia [512]. Similar results were obtained when *Methylocystis parvus* OBBP

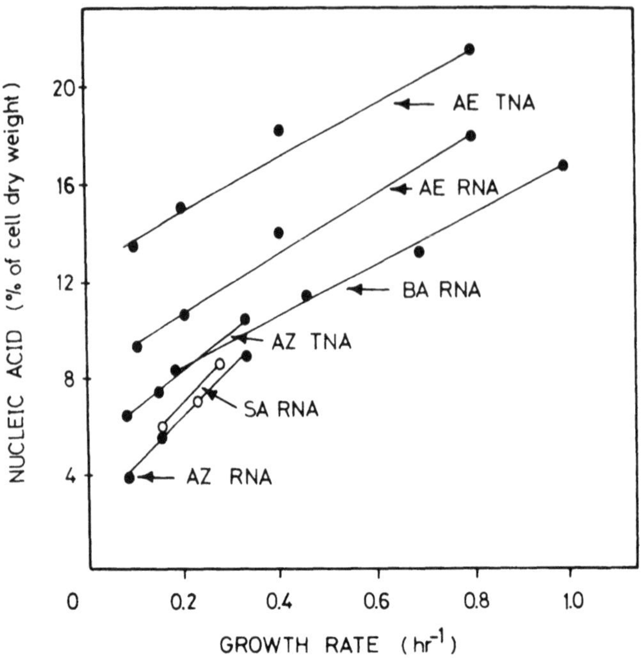

Fig. 30. Nucleic acid contents of cells as a function of growth rates. RNA, Ribonucleic acid; TNA, total nucleic acids; *AZ, Azotobacter vinelandii; SA, Saccharomyces cerevisiae; AE, Aerobacter aerogenes; BA, Bacillus megaterium.* (Redrawn from Mateles [723]). (Copyright © 1968. The MIT Press)

[732], and *Hyphomicrobium* sp. [733] were grown on methanol and produced extracellular viscous heteropolysaccharides. The reason for the difference between the yeasts and bacteria is not clear, but may be attributed to a slower rate of bacterial growth on nitrate. This hypothesis is difficult to check because of the problem of measuring growth rate in the presence of the large amount of polysaccharide produced. In another methylotroph, *Methylococcus* NCIB 11083, accumulation of intracellular glycogen/amylopectin occurred when growth on methane was limited by ammonia [734].

The bulk of the nucleic acid in microorganisms is RNA (e.g., more than 92% of the nuclei acid in *C. utilis* is RNA; see Dean [471]), which has a critical role in protein synthesis. Therefore, it may be anticipated that under carbon limitation the faster the rate of protein synthesis in a particular organism, i.e., at high growth rates, the higher the nucleic acid content. Figure 30 indicates that cells widely varying in nucleic acid content can be obtained simply by culturing them at different growth rates.

The nucleic acid content is affected both by the growth rate and by the temperature of growth [735–740]. A general approach for describing these changes was given by Alroy and Tannenbaum [731] who considered a plot of the RNA: protein ratio of various microorganisms against a modified growth rate (or dilution rate of a continuous culture). This modification was obtained by taking the

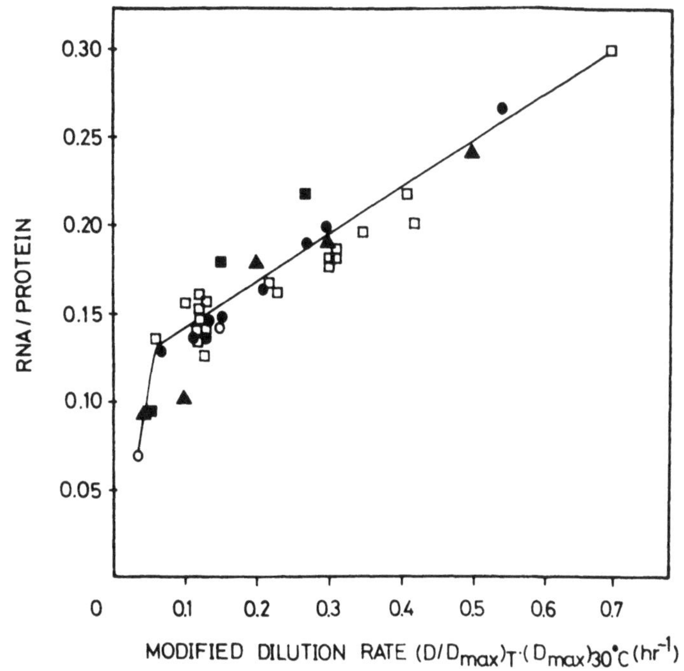

Fig. 31. RNA: Protein ratio as a function of modified dilution rate in a carbon-limited chemostat; ●, *Aerobacter aerogenes*; □. *Candida utilis*; ▲, *Bacillus megaterium*; ○, *Polytomella caeca*; ■, *Azotobacter chrococcum*. (Redrawn from Alroy and Tannenbaum [731]). (Reprinted with permission of John Wiley & Sons, Inc.)

ratio of the actual dilution rate (D) at a particular temperature (T) to the maximum specific dilution rate (D_{max}) at that temperature, and multiplying the product by the maximum specific dilution rate at 30 °C. This modification resulted in data presented in Fig. 31. This shows that the ratio of total nucleic acid to protein increased with an increase in dilution rate at constant temperature similar to results obtained by other works [738, 741, 742] and decreased with an increase in the temperature at a specific dilution rate. It is also suggested that the ratio of RNA to protein was approximately the same for various microorganisms cultivated in the chemostat under carbon limitation. However, the validity of this similarity at other cultivation temperatures was uncertain, since the activation energies for maximum specific growth rates of microorganisms were not identical [731]. Using the approach described above, it is possible to predict, fairly accurately, the content of nucleic acids to be expected for a particular organism considered for SCP production.

Table 28 shows that microorganisms contain relatively high amounts of nucleic acids, even when expressed on a protein basis, as compared with conventional foods of animal or plant origin. The nucleic acid content of the SCP product, mostly RNA, can vary from 8 to 25 grams per 100 grams of protein [630]. According to several clinical studies [702, 753] and to the recommendations of the Protein Advisory Group (PAG) of the United Nations (see the revised guidelines in [573]), safe daily intake of nucleic acids for most adult persons is estimated to be about 4 g, of which 2 g can be derived from SCP [5]. This amount of nucleic acids in the diet, corresponds to a nucleic acid content of 3% (of dry wt) in the SCP product if considerable amounts of SCP are contained within the diet. As the usual nucleic acid content in SCP is much higher, it is clear that in all cases use of SCP as a major component in human diet will require treatment to lower the nucleic acid content.

So, considerable attention has been devoted to the reduction of nucleic acid levels in the SCP product, and this accounts for the voluminous literature which deal with this important question (see the reviews by Kihlberg [31], Hedenskog and Mogren [744], Sinskey and Tannenbaum [703], Viikari and Linko [630], and Solomons [8]).

The approaches suggested to reduce the nucleic acid content of biomass preparations can be divided into two groups. Those which aim to produce biomass with a low nucleic acid content and those which aim at the reduction of RNA content of the SCP product in the downstream processing. Obviously, it may seem ideal to obtain biomass with lower RNA content by suitable control of the fermentation process (dilution rate, temperature, phosphate limitation, etc.). However, due to the fact that protein synthesis is tightly bound to RNA synthesis, only a limited decrease in RNA content is possible without concomitant decrease in protein content.

The other approach aims for the reduction of the nucleic acid content of the SCP product by removal of nucleic acids via a variety of treatments listed in Table 32. When considering these suggested methods, special attention should be paid to the safety and economics of the process. Some of the methods seem to be complicated, expensive and hardly applicable at an industrial scale [8, 630].

A simple and rapid method to reduce the nucleic acid content of the SCP product is the base-catalyzed hydrolysis by NH_4OH, KOH or NaOH. It is due to the high susceptibility of RNA to base hydrolysis, and the SCP obtained after this treatment can be textured immediately. The alkaline treatment markedly improved the consistency, odor and taste of the biomass when neutralized. One disadvantage includes the losses of weight, as the supernate obtained after the alkaline treatment contained not only the partially hydrolyzed RNA, but also protein molecules.

Other chemical extraction procedures may be employed such as treating the dried cell mass with phenol, anhydrous NH_3 and methanol or with sodium chloride (Table 32). Although this method is relatively simple, rapid and removes polymerized RNA from whole cells, it is counteracted by the loss of amino acids and low molecular materials such as vitamins [703].

It has been suggested that cell disruption would be necessary to increase the digestibility of the cells. The disruption of cells can be done mechanically, enzymatically or chemically. The RNA obtained is enzymatically, chemically or chromatographically removed from cell extract (Table 32). In general, these procedures are rather expensive and therefore will be applied only if the protein isolates or concentrates are used for human consumption.

Exogenous hydrolyzing enzymes may also be used to deplete SCP products of nucleic acids (Table 32). This simple method permits milder conditions than

Table 32. Methods for RNA reduction in the SCP product

Method	Protocol	Organism	Initial and final %RNA (of dry wt)	Ref.
Base-catalyzed hydrolysis	0.12 N NaOH for 30 min at 50 °C	*P. variotii*	9%→3%	[630]
Chemical extraction	5% NaCl at 120 °C	Baker's yeast	Complete removal	[743]
Cell disruption; physical separation:	Disintegration-glass beads	Different microorganisms	Final % RNA = 2	[744]
a) Enzymatic treatment	Malt sprout nuclease	Yeast	Final % RNA < 3	[745]
b) Chemical	Succinic anhydride pH 4.2–4.4	*S. carlsbergensis*	6–8%→1.8%	[746]
c) Chromatography	Cellex E-Ecteola cellulose	*C. utilis* *S. cerevisiae* *S. carlsbergensis* *Z. mobilis*	Less than 3% of initial content	[747]
Exogenous RNAase	Bovine pancreatic ribonuclease 55–56 °C, pH 6.7–8.0	*C. utilis* *C. intermedia*	Final % RNA = 1.5	[748]
Endogenous RNAase – heat shock	90 °C, pH 2, 20 min	*R. glutinis*	6.5%→1.2%	[749]
	45 °C, pH 5.8–6.9, 2 h	*C. utilis*	6–8%→2%	[750]
	68 °C for a few sec 50 °C 1 h, 55 °C 1 h	*C. utilis*	7–8%→1–2%	[751]
	50–60 °C + NaCl	*S. cerevisiae*	7–8%→1.4%	[752]

chemical methods. Disadvantages of this method include the pretreatment requirement of cells to increase permeability, thus increasing cost, high weight loss (reducing process yield) (as do all methods) and the high cost of enzyme (the exogenous nuclease preparation should be free of proteases).

Several reports have demonstrated that cellular endogenous ribonucleases may be activated by heat treatment or by the use of chemicals (Table 32). This activation which was first described by Maul [754], may occur by changes in cellular location of ribonuclease, by conformational changes in ribonuclease itself, or by denaturation of ribonuclease inhibitors [630]. The biochemical mechanism of activation seems to vary between different microorganisms [630], and no correlation was found between endogenous enzyme levels and the effectiveness of the heat-shock process for reducing RNA content [703]. Disadvantages of this method include weight loss, long duration of the heat treatment, and effectiveness only on certain cells.

Sinskey and Tannenbaum [703] employed one yeast strain, *Candida utilis* Y-900, in their comparative study on the various methods used for lowering the nucleic acid content in the SCP product. Alkali treatment of whole cells resulted in abnormally high yield, probably due, in part, to the excessive precipitation of materials other than cells, plus the fact that a higher salt content resulted from the NaOH used. Dry weight yields were similar for all the other processes, with a low value of 67% for exogenous enzyme treatment to a high value of 82% for the heat shock (endogenous enzyme) process. The last two processes resulted in the highest yield of protein (84%) with a final nucleic acid content of approximately 1%.

3.3.1.2 Lipids

Environmental factors, e.g., temperature and composition of the media, are known to affect lipid content and composition in diverse microorganisms [755]. In this section, several studies on the responses of prokaryotes and eukaryotes to environmental changes are described.

The efficiency of lipid production by the fungi *Tolyposporium ehrenbergii* and *Sphacelotheca reiliana* cultivated on media containing various mono- or disaccharides has been studied [756]. Sugar cane molasses, potato starch, glucose syrup and glycerol, were used as media ingredients. Media containing fructose or glucose and peptone were the most efficient for lipid production during a one week incubation. After a two week incubation period, highest phospholipid contents occurred in mycelia obtained from media containing molasses and peptone, or glycerol and urea. These results indicated that both the content and composition of lipids were affected by both the culture medium composition and the age of the culture.

The efficiency of lipid biosynthesis by 16 yeast strains growing on ethanol was studied [757, 758]. The yeast strains were from the genera *Cryptococcus, Torulopsis, Candida, Lipomyces, Zygolipomyces, Trichosporon,* and *Endomyces.* The yeast lipid content varied from 20 to 66% of dry weight, and lipid accumulation was most pronounced in strains with low protein content. In another study, it was shown that the carbon and energy source for aerobically grown cultures of *Candida guilliermondii* profoundly influenced the neutral lipid content and the fatty acid composition of the individual lipid components [759]. Methanol-grown yeast contained 12% total lipids, 67% of which were neutral lipids. Cells grown on glucose or ethanol contained 21 to 22% total lipids, 80% of which were neutral lipids. Methanol-grown yeast contained 52 to 54% of 18:1 acid (of total fatty acids) and 23 to 25% of 18:2 acid, as compared with yeast cells grown on glucose or ethanol which contained 57 to 66% of 18:1 acid and 8 to 14% of 18:2 acid, in both neutral and polar lipid fractions [759]. Different results were obtained with the methanol-utilizing bacteria *Methylophilus methylotrophus, Pseudomonas extorquens, Microcyclus aquatica* and *Hyphomicrobium polymorphum* [698, 760]. Cellular fatty acid composition of these strains did not change when the bacteria were grown on media containing methanol, peptone or nutrient broth at 30 °C, 37 °C, and 42 °C. The fatty acid composition of the bacterium *Pseudomonas* C was similar when it was grown on methanol in a batch culture, and in a continuous culture at different dilution rates [697]. These results indicated that bacteria are less susceptible to changes in their fatty acid composition than yeasts.

Aliphatic alkanes undergo oxidation by way of an aldehyde, yielding a fatty acid of identical carbon number to the alkane substrate [210, 761, 762]. Cellular fatty acid composition of the yeasts *Candida tropicalis* and *Candida lipolytica* were studied during growth on individual n-alkanes of various chain lengths (from C_{11} to C_{18}), and a mixture of n-alkanes (C_{10} to C_{13}) as the sole carbon sources [763]. Distinct relationships between the chain length of n-alkane substrate, and the patterns of the fatty acid compositions of these yeasts, especially those of the cells grown on odd-carbon alkanes were noted. The data obtained indicated that the accumulation of the fatty acid having the same chain-length as that of the alkane substrate used, was predominant in the yeasts cultivated on a longer alkane such as n-heptadecane and n-octadecane. On the other hand, the "chain elongation pathway" and the "de novo synthesis pathway" following β-oxidation of substrate were simultaneously operative in cells growing on a relatively shorter alkane, such as undecane and dodecane.

A comparison of the lipid content and fatty acid composition of yeasts grown on glucose and hydrocarbons (n-alkane mixture) was made [763]. The content of total lipids in glucose-grown *C. tropicalis* cells was 6.9%, while that in the hydrocarbon-grown cells was 12% [763]. A significant difference was observed between the proportions of odd-carbon acids, in particular, C_{17}, in glucose-grown cells and hydrocarbon-grown cells; 48.9% odd-carbon fatty acid (of total fatty acids) was detected in hydrocarbon-grown cells, as compared to 6.6% in glucose-grown cells.

Different relationships between hydrocarbon oxidation and fatty acid metabolism were reported for various bacteria. Raymond and Davis [764] observed a marked increase in the lipid content of a *Nocardia* species grown on n-alkanes. Cells grown on n-hexadecane and on n-octadecane contained about 70% lipid material. In contrast, cells grown on n-hexane and n-tridecane contained about the same amount of lipids as glucose-grown cells. Davis [765] grew a *Nocardia* species on n-alkanes with 13 to 20 carbon atoms. He found fatty acids of cellular triglycerides and aliphatic waxes with carbon skeletons, analogous to the carbon skeleton of the n-alkane substrate on which the organism was grown. Similar findings were observed in cellular lipids of a *Mycobacterium* grown on n-alkanes with 13 to 17 carbon atoms [766]. In *Micrococcus cerificans* the phospholipid fatty acid chain length directly reflected the carbon number of the alkane substrate, with 40, 84, 98, and 77% of the fatty acids being 14, 15, 16, and 17 carbons when cells were grown on C_{14}, C_{15}, C_{16}, and C_{17} n-alkanes, respectively [236]. An increase in total unsaturated fatty acids localized in the phospholipids was noted in cells grown on alkanes greater than 15 carbons in length. An extracellular accumulation of free fatty acid was demonstrated in hexadecane-grown cultures and was not apparent in non-hydrocarbon-grown cultures.

The above cited studies showed that in these bacteria, or yeasts, de novo fatty acid biosynthesis (from acetate) was inhibited, and the cells were able to incorporate exogenous fatty acids and maintain proper membrane structural characteristics with respect to alkane chain length. However, other bacteria showed only a partial correlation between substrate chain length and chain length of cellular fatty acids [766–769]. Thus, the total lipids produced by *Pseudomonas aeruginosa*, growing on a pure hydrocarbon (tridecane) or mixed hydrocarbons (JP-4 jet fuel), represented 7.1 to 8.1% of the cellular dry wt as compared to 8.2% obtained in cells grown on Trypticase Soy Broth [770]. Quantitative differences were found in fatty acids produced on these media, but specific correlations between substrate carbon sources and fatty acid content of cells were not recorded. Tridecane-grown cells contained only traces of C_{13} acid and small amount of C_{15} and C_{17}

acids, suggesting that the organism's fatty acids were derived from de novo synthesis rather than by direct incorporation of the hydrocarbon.

3.3.1.3 *Carbohydrates*

Another aspect of changing the macromolecular composition of the SCP product concerns human food to be used in extended space missions. It is estimated that an ideal space diet contains 70% carbohydrate [771]. Microbial biomass is generally compatible with human food requirements but lacks adequate carbohydrate levels; thus, attempts were made to increase the levels of storage carbohydrates, principally glycogen, through cultural and genetic methods [772]. A methylotrophic yeast, *Hansenula polymorpha* DL-1, was chosen in order to fit the needs of a hybrid chemical-biological scheme for food generation in space. The effect of nitrogen (as ammonium sulfate) limitation, and 0.02 M potassium concentration, resulted in increasing the storage carbohydrate content of the cell dry weight from 30 to 46%. Several mutant strains were obtained, and they were selected by the iodine staining technique previously used for the isolation of glycogen-deficient and high glycogen mutants. In the stationary phase the isolated mutant strains exhibited an additional 4 to 16% increase in the levels of storage carbohydrates over the parent strain. The combined effect of media formulation, mutation and selection produced yeast strains that could accumulate carbohydrates up to 60% of the dry weight, with a net gain in glycogen of 4 to 16% over the parent strain. The approach taken with this particular yeast has, in the author's opinion, generalized utility, and could be used to increase the carbohydrate content in any candidate microbial species to a level closer to those of the human dietary requirements in space.

3.3.2 Environmental and Genetic Control of Protein Composition

The potential economic significance of obtaining SCP with an increased level of some amino acids, particulary methionine and cysteine, requires serious consideration of various strategies for altering the amino acid pattern of SCP. Before considering these strategies it should be mentioned that deficiencies in sulfur amino acids can be overcome by blending and enrichment, but these are not necessarily practical measures. There is also the possibility of causing extreme over-production of one or more amino acids which are excreted to the fermentation broth; this process is, in fact, the basis for a large industry for amino acid manufacturing. However, these amino acids are found in the free amino acid pool and are excreted to the medium, and thereby are lost during harvesting operations of the biomass. A more fundamental and lasting solution would be physiological adaptation or genetic modification of the microorganism of choice to produce protein with a better amino acid balance. Alternatively, if the free amino acid pool could be enriched in a nutritionally significant amino acid, and if the amino acid could be rendered insoluble or polymerized during or after the production stage, the resulting SCP would be more desirable [36]. While some consideration was given to the first approach, there are no publications dealing with the latter one.

3.3.2.1 Environmental Conditions

Several studies performed with *Enterobacter aerogenes* [773], *Escherichia coli* [773], *Azotobacter agile* [774] and *Saccharomyces cerevisiae* [775] indicated that the amino acid composition of total cell protein is remarkably constant with respect to such cultivation variables as medium composition, pH and temperature. A similar observation was made when the unicellular algae *Chlorella ellipsoidea* was subjected to alternate periods of light and darkness [776]. In other microorganisms, however, changes in physiological conditions affected their amino acid profiles. Thus, when growth phase was the independent variable with *Azotobacter agile*, variations occurred in the basic amino acids: lysine, arginine, and histidine. The levels of these amino acids, relative to the sum of all amino acids, appeared to increase from 45 to 65% as the culture advanced from lag to log phase and, ultimately, to decrease to their initial values when the culture approached stationary phase [774]. The authors suggested that the variations were related to changes in the nucleoprotein content of the cells. In a somewhat similar study, Seoka [777] found a noticeable increase (9 and 17%, respectively) in the relative levels of lysine and arginine in *E. coli*, when log phase cells were compared with stationary phase cells. Malofeeva and Belyanova [778], showed that by shifting the culture of *Chloropseudomonas ethylica* from growth on CO_2 in light to acetate utilization in total darkness, the ratio of change for the respective amino acids was: histidine, 1.59; threonine 1,19; valine, 1.16, and tyrosine, 1.28. The authors proposed that these variations reflected a major change in protein composition, which was caused by the disappearance of the photosynthesizing apparatus. In other detailed studies [223, 779, 780], a small difference was noted between the amino acid composition of *Candida tropicalis* N_7Y_1 cells grown on hydrocarbons and on carbohydrates. A hydrocarbon-containing medium yielded more cystine than a carbohydrate-containing one. However, when *Candida tropicalis* N_7Y_1, or *Pseudomonas aeruginosa* S_7B_1 were grown on n-hexadecane at different pH's, the amount of essential amino acids varied considerably. Thus, when the growth of *P. aeruginosa* was carried out at pH 7.0 and pH 8.0, the respective content of lysine and methionine varied from 2.88 to 6.45% and 0.79 to 1.29% of total amino acids. In *C. tropicalis* the content of lysine at pH 6.0 was 4.07%, while at pH 8.0 it was 1.92% of total amino acids. (It should be noted that due to the vast differences in processing techniques, comparisons of this nature have limited value.)

DeZeeuw [781] maintained that environmental changes appeared to offer little potential for significantly modifying the amino acid profile of microbial protein. This conclusion was rationalized on the basis of two considerations. First, proteins with amino acid contents at great variance from the normal are not produced in large proportions. Second, the environmental changes led to accomodations over a major portion of the enzyme profile and not of just a few proteins, and consequently tended to average out the amino acid content [781].

The possible effects of changes in the levels of individual proteins on the amino acid composition of total cell protein have been explored theoretically and experimentally by Alroy and Tannenbaum [782, 783]. In their first paper they presented a mathematical model which described the concentration of an amino acid, in total protein, as a function of its concentration in individual cell proteins or in sets of cell proteins [782]. The resulting equation made it possible to calculate how the make-up of cell proteins must change in order to obtain a specified alteration in the content of a specific amino acid. On the basis of this analysis, Alroy and Tannenbaum [782], concluded that the amino acid composition of total protein is not likely to be significantly affected by variations in the cultivation conditions, unless there exist a set of proteins which are related by composition and function. They suggested that ribosomal proteins, all of which are probably moderately rich in basic amino acids and whose most important function is to interact with acidic substances (nucleic acids), might constitute such a group of proteins. In their second paper, they examined changes in amino acid content resulting from the growth of *Candida utilis* and *Enterobacter aerogenes* under different conditions, namely, various pH levels, temperatures, dilution rates, nitrogen sources and nitrogen and phosphorus limitations [783]. Variations in the content of some amino acids, e. g., lysine and arginine, could be correlated with growth rate and were explained by changes in the proportional amount of ribosomal protein at the different growth rates. Thus, the increase in lysine and arginine content was due only to an increase in the levels of the ribosomal proteins, at higher growth rates, relative to total cell protein, whilst the amino acid composition of the ribosomal protein fraction, or the total cell protein less the ribosomal protein, remained nearly constant. The authors argued that although an increase in the ratio of ribosomes to cell protein would be obtained if the growth rate were accelerated, this increase would be limited and, therefore, only small to moderate variations in the relative levels of specific amino acids would be expected [783].

3.3.2.2 Genetic Control of Protein Composition

DeZeeuw [781] argued that since some proteins must possess a higher-than-normal methionine content, the opportunities for increasing a single protein's contribution to the total, need to be considered. (It is interesting to note that such proteins do exist, e. g., recently, research at ARCO-PCRI has led to the cloning of a Brazil-nut methionine-rich protein-17% methionine content, 28% total sulfated amino acids [784].) He suggested that under suitable conditions, the protein of a given inducible enzyme could represent 5 to 10% of the total cell protein. Thus, a single enzyme could initiate a change, although relatively small, in the amino acid balance of microbial protein. The level of methionine in the specific protein could then be increased by obtaining mutants in which base transitions and transversions (point mutation) effected the substitution of methionine for other amino acids. However, at that time there did not appear to be a rationale for selecting such mutants. Only later (see below), when recombinant DNA techniques became familar did other scientists begin to examine this idea.

Okanishi and Gregory [785] tried to improve the amino acid composition of SCP by isolating yeast mutants which formed smaller-sized colonies than their parent, when grown on a sulfur-deficient medium. This approach was based on the assumption that methionine-rich mutants would have a higher sulfur requirement for growth than their wild-type parent. In this way, mutants of *Candida tropicalis* were isolated that contained up to 41% more methionine than the parent, on a dry-weight basis. It was postulated that the mutants were derepressed,

i. e., some protein that was rich in methionine was produced. However, the parent culture was particularly low in methionine, and even the best mutant still had a suboptimal amount of methionine. Moreover, Momose and Gregory [786] questioned these results by arguing that a derepressed protein containing 5% methionine would have to form more than 70% of the total proteins, whereas a highly methionine-rich protein containing 10% methionine would have to constitute more than 30% of the total proteins.

A novel approach to the problem has been proposed by Momose and Gregory [786]. They reasoned that it should be possible to isolate mutants with temperature-sensitive mutations in either aminoacyl-transfer ribonucleic acid synthetase, transfer ribonucleic acids, or the ribosome itself. At a nonpermissive temperature the microorganism would begin to synthesize all of its proteins, with methionine substituting in part for a more common amino acid (s). The expression of such mutations could be lethal to the microorganisms, depending on the degree of alteration of the amino acid pattern of essential proteins, whereas conditional mutations would allow the microorganism to grow normally at the permissive condition. Furthermore, errors resulting in the incorporation of methionine in place of the correct amino acid might be expected to occur more readily in the presence of a high concentration of methionine. Therefore, temperature-sensitive mutants, with an increased sensitivity to excess methionine at an elevated temperature, were considered to be possible candidates for the type of mutant sought. Temperature-sensitive mutants were derived from *Saccharomyces cerevisiae* Y5α by ethyl methane sulfonate mutagenesis. Most of the temperature-sensitive mutants isolated (out of a total of 132 mutant strains) had the same methionine content as the wild-type strain. However, several of the mutants were found to increase the proportion of methionine in their protein to much higher levels than that of the wild-type parent after a temperature shift from 25 to 37 °C. Two strains, 476 and 438, which were temperature sensitive only in the presence of methionine (Ts-Met^s(p)), produced cellular protein with methionine contents as high as 3.6 and 4.3%, respectively, when incubated in the presence of methionine, as compared to the wild-type strain which had 1.75% methionine under all the conditions tested.

The mechanisms of variability in methionene content in the total cellular proteins are unknown. Momose and Gregory [786] suggested that the plausible explanation is that protein synthesis was altered in a way that methionine competed with one or more amino acids for incorporation.

It should be pointed out that since the methionine assays were performed on hydrolyzates of whole cells, it has not been definitively established that the increased cellular methionine was incorporated into protein [786]. Although the non-permissive conditions would have to be obtained subsequent to growth, for instance by running a two-stage continuous fermentation with second stage held at the non-permissive temperature, the concept of this research is of considerable interest and bears further examination [36].

Recently, several mutants of a methanol-assimilating yeast, *Pichia pastoris*, were isolated which contained a higher concentration of methionie [787, 788]. These mutants were selected after ethylmethane sulfonate treatment, based on their resistance to the antimetabolites ethionine and norleucine, and on their ability to cross-feed a strain of *Bacillus subtilis* that requires methionine for growth. The methionine content as percent total protein in these mutants ranged up to 6%, compared to 1.2% for the parent strain. There are indications that methionine was included in protein and not excreted to the culture broth; coculture with *B. utilis* auxotroph for methionine did not result in the growth of the bacterium. The protein content and productivity of the yeast mutant were similar to the parent strain when grown in a chemostat [787].

Due to recent developments in recombinant-DNA techniques, it was possible for Gautier and Bonewald [789] to reexamine the idea previously suggested by DeZeeuw [781] for improving the amino acid composition of SCP. They have suggested that it might be helpful to express foreign genes with high amounts of the desired amino acid in the gene product within methanol-utilizing bacteria, or alternatively the genes for methanol-utilization could be transferred to other appropriate bacteria. Recognizing the limited, yet, important change in amino acid composition which could be obtained by this method, they have demonstrated the first step in the proposed scheme. They described a cloning system for the methanol-utilizing *Pseudomonas* AM-1 and *Pseudomonas* 3ab. In this system they used plasmid derivatives of Rll62 which were constructed by in vitro manipulation, and could replicate in *E. coli* and *Pseudomonas*. As plasmid Rll62 could be mobilized to *Pseudomonas* AM-l with high efficiency, transformation and hybrid plasmid selection could be done within *E. coli*, followed by mobilization of the hybrid plasmids into methanol-utilizing bacteria.

This approach was slightly modified by researchers at M.I.T. [790]. They reasoned that the composition of a protein may be enriched for a given amino acid by insertion of a synthetic polynucleotide sequence through genetic engineering. This was successfully demonstrated by modifying the ampicillinase protein of pBR322, by insertion of poly G-C sequences, resulting in an increase in proline content of the ampicillinase protein in *Escherichia coli* [790].

Undoubtedly, this intriguing subject will receive greater interest in the near future.

3.4 Toxicological Status and Nutritional Value of SCP Products

Before any SCP product can be used in animal feeds, let alone as a human dietary constituent, it is essential to undertake extensive tests on animals to determine the toxicological state of the product and its nutritional value. The usual toxicity testing methods of foodstuffs are applied mainly to food additives, which are added in small quantities to the food. These methods require some modifications when applied to SCP as its products are intended to be used as major components of diets. Single cell protein products differ one from another; thus each indiviual SCP product must receive its own safety approval.

In order to carry out toxicological studies on the SCP, a sufficient amount of microbial material must be produced in a pilot plant. If these tests are successful, a provisional approval is given to the producer, and the construction of a production plant can be undertaken. Once a single cell protein product is commercially manufactured it is necessary to requalify the material as being identical with that previously cleared for pilot plant production [791]. This procedure requires considerable time and risk on the part of the producer, together with a large expenditure for animal testing. For example, ICI used approximately 500 tons of "Pruteen" in feeding 200,000 animals for seven different species at a cost of approximately $ 6 million [8]. Due to government regulations, and the concern of the producer to sell a new and safe food or feed component, the test programs that have been pursued during the development of SCP have been by far the most comprehensive ever devised for foods. Obviously, only a limited number of companies, having those financial resources, can carry out such extensive safety evaluation programs to test adequately their SCP products.

In order to provide information and suggest sensible protocols for testing of SCP products for both human food and animal feed, the Protein Advisory Group (PAG) of FAO/WHO/UNICEF issued a series of guidelines. The nutritional and toxicological testings specified by these guidelines involve short-term studies for acute toxicity, long-term life-span and multigeneration studies, teratological and mutagenic studies, and sensitization and immunoresponse studies [8].

The potential toxicity of SCP products can be considered through knowledge of the nature of the organism(s) grown in the fermentor (including the contaminating organisms), the types of metabolites they produce, and the residues present as a result of the specific fermentation process [793].

Obviously, all the microorganisms that are used for SCP production are known to be nonpathogenic organisms. Nevertheless, concern has been expressed that a microbial strain used in this process could undergo mutational changes, resulting in its development of toxins or other nutritionally adverse characteristics. However, long-term continuous fermentation for production of SCP on pilot plant or industrial scale has revealed the remarkable stability of the strains. Another concern was with regard to the impact that process variations would have upon the toxicological characteristics of the product. This was studied in the case of ICI's "Pruteen" [793]. Variations in the production mode of *Methylophilus methylotrophus* on methanol were envisaged in the fermentation (including the sterilization methods used), and in the modes of harvesting and drying of the SCP product. Microbial biomass samples were produced on a pilot scale, i.e., in an air lift fermentor of a 40-m^3 working volume, which reflected these modes of operation, and were compared in various tests involving both laboratory animals (mice, rats, and dogs), and farm animals (chicken, fish, and pigs). It was found that process variations did not induce any biological variations that would indicate formation of a toxic material [793].

Another problem to consider is the possibility of contamination by pathogenic microorganisms. Although the downstream processes of SCP products are designed to destroy most of the viable microorganisms present, some could survive in the product. Recommendations have been established for microbiological standards of SCP products for use in animal feeds (PAG guideline No. 12), and even though these were very restrictive, several SCP products did show a low contamination level (less than 10^5 organisms per gram) and did not contain organisms of major public health significance [8, 205, 689].

The SCP product should be free of toxic compounds (or metabolites) produced by the microorganisms(s). Several of these potentially toxic metabolites were considered by some experts.

1) Nucleic acids – their content in SCP products and the toxic effects they cause in humans were discussed earlier in this chapter. Also, several methods have been mentioned (Table 32) for lowering the nucleic acid content of microbial biomass so that safe application of this protein source for human consumption is possible.

2) Mycotoxins – these extremely toxic secondary metabolites are produced by many types of filamentous fungi. Therefore, any fungus which is proposed as a microbial food or feed is subjected to very close scrutiny for possible mycotoxin production [12].

3) Endotoxin – this is a component associated with the outermost lipopolysaccharide layer of many Gram-negative bacteria. Little has been published concerning the possibility that the presence of this toxin will make the bacterial biomass unacceptable for animal consumption.

4) Uneven-numbered fatty acids – these occur in the lipids of yeasts grown on uneven-numbered n-paraffins, i.e., C_{13}, C_{15}, and C_{17}. Although such fatty acids are normally found in a number of foods in varying quantities, it was suggested that in hydrocarbon-grown yeasts these acids are present at relatively high quantity and might cause adverse effects when fed to animals [722].

5) Other fatty acids – a relatively high content of palmitoleic acid ($C_{16:1}$ – the normally occuring fatty acid in butter) has been found in SCP products prepared from some methanol-grown bacteria (Table 31). These cells also contain cyclopropane fatty acids [793]. So far, no experimental evidence has been obtained to show that all these fatty acids are harmful to the physiology of the test animals. It has been shown that they are utilized according to known physiological metabolic pathways.

6) D-Amino acids – these are part of the cell wall of bacteria and limited to D-glutamic acid and D-alanine [793]. They are known to be of no toxicological consequence, due to their low content in the bacterial SCP product [793].

Toxicity of the SCP product may arise not only from a microbial component or metabolite, but from unused nutrients. Thus, the fear was expressed that residual hydrocarbon fractions from gas-oil might be a hazard to animal health, especially because of the possibility of the carcinogenic residues remaining in the product. The fear appears to be unjustified, at least for several types of SCP products, i.e., "Toprina", "Liquipron," and "Kanepron." This is because the substrates used were highly purified and the content of polycyclic aromatic compounds in the hydrocarbons-based SCP product, especially that of 3,4-benzpyrene, was reported to be below 1 ppb (μg/kg), whereas that of baker's yeast has been reported to be 12 to 13 times as much [8]. In addition, multigeneration studies on different animals did not reveal any carcinogenic or teratogenic response (see below).

A nephrotoxic factor has been found in alkali treated soybean protein and identified as lysino-alanine [794]. It was shown that in the case of "Pruteen" the formation of lysino-alanine is not a problem and at no stage does the process approach pH 10, a necessity for the formation of this toxic amino acid [793].

Microorganisms assimilate and condensate some heavy metals from the medium. However, it appears that control over the process raw materials ensures that the level of toxic heavy metals is not exceeded (Table 30).

Shacklady [721] has published an extensive range of toxicological and nutritional test results on "Toprina," a product of *Candida lipolytica*, produced by British Petroleum both at Grangemouth, Scotland and at Lavera, France, on n-paraffins or on gas-oil. The test protocols included acute toxicity, subchronic toxicity, chronic toxicity, and carcinogenicity. These tests were performed on rats and mice for up to two years (equal to their life span) with 10, 20, 30, and 40% of the diet as "Toprina." Multigeneration tests were performed on rats and Japanese quails for 15 and 23 generations, respectively. In addition, teratogenicity studies, mutagenicity studies, sensitization studies, and induced toxicity studies on products from pig and poultry fed on alkane yeasts were performed.

Series of single cell proteins were evaluated, in accordance with the criteria specified by PAG guidelines. Examples are *Candida maltosa* "Kanepron" product by Kanegafuchi [33, 795], *Candida paraffinica* "Viton" product by Dainippon (Dainippon Ink and Chemicals, Inc. service report, undated, personal), *Methylophilus methylotrophus* "Pruteen" product by ICI [689], and *Fusarium graminearum* "Mycoprotein" product by RHM [8]. These products have been manufactured on different scales, and no hazards were detected in the experimental work.

In subsequent commercialization of SCP products, the SCP intended for human use naturally necessitated careful testing in humans in addition to extensive animal testing. Currently, only a few SCP products are permitted to be used as human food. In the USA, the Food and Drug Administration (FDA) regulations permit the human use of dried cells of *S. cerevisiae* (baker's yeast) *C. utilis* (Torula yeast), *K. fragilis* (Fragilis yeast) and baker's yeast protein concentrate. In the U.K., the Ministry of Agriculture, Fisheries, and Food has allowed test market studies on the food use of mycelia of *F. graninearum* developed by RHM [7]. Information is not available on SCP products permitted as food supplement in the Eastern European countries.

The nutritional quality of SCP is generally determined by the estimation of digestibility, biological value, and either net protein utilization [2] or protein efficiency ratio [3] in animals, and by nitrogen balance studies in man (PAG Guideline No. 12). It was pointed out that "ideally, the nutritional evaluation of protein foods should be made in relation to their potential role in the diet of the population for whom they are intended" (PAG Guideline No. 6). However, the usual manner of evaluating the quality of various SCP products (like other food additives) is by feeding tests with experimental animals, of which the rat is the most

2 Net protein utilization (NPU) – by this method the total content of nitrogen in the body of the animal of both the protein-fed and protein-free groups is determined

3 Protein efficiency ratio (PER) – by this method the average net gain in body weight per unit weight of protein ($N \times 6.25$) is compared with that observed for the reference protein, casein. Both these methods (2 and 3) utilize rats which are fed with the diets for 10-day or 28-day periods (PAG Guideline No. 6)

widely used species. In these tests, careful formulation of the test diet is required, because SCP is added at high levels and might alter the nutritional balance of the diet, bearing false results [8]. Indeed, modern animal feeds are blended by the aid of a computer from selected batches of their constituents of cereals, soybean, fish-meal, etc., in such a way that the composition and especially the crude protein content of the final product is kept constant with low variance [701].

In competing with other foodstuffs, SCP must have a high content of species nutrients (mainly essential amino acids) and a high energy content in relation to its price. The specific nutrients, furthermore must, be well balanced with, or complement to, other feedstuffs in the diet [1976]. In an animal feed, SCP would score because of its consistent quality, which in modern compounding is paramount importance [624].

For appraising their nutritive property, several SCP products have been fed to animal species, ranging from rodents to domestic livestock, and found to support growth and reproduction in these animals when used in substitution of natural protein supplements, such as soybean meal, fish-meal, and skim milk powder [5, 32, 50, 55, 216, 380, 457, 689, 694, 721, 797, 798].

In general, SCP (in range of 5 to 15% of the total rations) has a higher nutritional value than comparable quality soy proteins, and hence can have particular utility for fortifying food products. Microbial proteins are rich in lysine and relatively poor in methionine. Synthetic D, L-methionine, as it is rather inexpensive, can be added to improve the quality of the SCP, especially for its use in the diet of poultry.

Obviously, microbial biomass is nutritionally valuable to man and animals only if the protein and vitamins contained in the biomass are first digested by intestinal enzymes, and then absorbed and utilized by the cells. Concern was expressed that the cell walls of microorganisms, especially of algae [46], would prove to be a major barrier to the digestibility of the internal protein of the microbial biomass. It was found, however, that this did not prevent the use of microbial biomass as a food supplement. Nevertheless, on occasion it may be advisable to feed a microbial biomass that has been subjected to a treatment that releases protein from the cells (see Tables 23 and 32 [12]).

4 Concluding Remarks and Epilogue

During the last forty years, enthusiasm for biotechnology has fluctuated sinusoidally. Maxima were indicative of people's "absolute confidence" that "solutions to world problems" had been found in the field. Obvious examples were the discovery of antibiotics, development of an alternative protein source (SCP), enzyme immobilization for multiple use in catalyzing various processes, ethanol production from renewable resources and, of course, the tremendous scientific achievements in molecular biology obtained through the use of genetic engineering techniques. Contrary to these, minima were indicative of realistic acknowledgement of problems in implementation of these new processes.

Without delving into the history of biotechnology per se, we shall analyze a case of "technological optimism versus economic and market forces" [8], namely, factors influencing the production of SCP. It will then become apparent why the rapid proliferation of commercial SCP plants was not realized.

Single cell protein production was meant to provide an additional source of protein, and to supplement conventional human foods and animal feeds. In the introduction, the advantages of SCP over other proteins were expounded. Essentially, SCP production is climate and season independent, requiring minimal land area, harvest time, and labor input, yet yields a homogeneous, high quality protein product. Its continuous production can be highly automated and high yielding over long periods of time.

The main factors justifying the extensive research and development programs focused on different aspects of SCP production are specific economic considerations. For example, SCP was needed as a means of providing relatively low cost protein in many countries.

Local interest tends to boost the impact of SCP to the market, particularly in the case of Eastern Europe's defecit in foreign exchange, ICI's wanting to be the pioneers of this technology, and BP wishing to secure certain political advantages in West Africa.

Single cell protein has not proved to be the solution to hunger in developing countries (see below), nor has it generated good economic returns.

As a consequence of early publications in the field, extremely optimistic opinions were expressed which suggested that SCP would act as a solution to the world's hunger crisis (see the optimistic papers of the First International Conference on SCP at M.I.T. in 1967 [1]). However, an analysis of the present production of SCP worldwide shows that these hopes were too optimistic. Current annual production output of SCP is roughly two million tons, the majority of which is yeast from cane or beet molasses. This production value may be wrong in its assumption that the USSR produces more than a million tons annually [799–801], and Solomons [8] seriously questions this figure (see also the recent article by Zilinskas [802] on biotechnology in the USSR). An estimate of capital investment in research and development is thought to be around $ 1000 million according to Solomons [8], with several large companies responsible for the major part (BP, ICI, Shell, Liquichimia, Dainippon, Mitsubishi, RHM, etc.) and numerous research groups playing fractional parts.

The estimated world's protein shortage was about 10 million tons in 1980. The gap between supply and demand is forecasted to grow to 25 million tons for the year 2000, indicating that the contribution of SCP to narrow the protein gap is limited [803]. Reasons for failure to reach high production rates include:

1) increased substrate costs – oil-based feedstock costs rocketed in 1973,
2) decreased relative cost of alternative proteins – soybean protein costs not having risen in the same way that SCP production costs have,
3) highly increased energy costs – also due to 1973 oil price increase,
4) increased plant construction costs – relative to low capital expenditure in classical agricultural projects,
5) increased interest rates in Western countries,
6) political directives,
7) psychological barriers of public opinion in accepting a new product – led to costly media struggles between consumer groups, government disapproval, and the producers (especially in Japan and Italy),
8) difficulty in providing a tasty product with no digestion problems,
9) lack of experienced manpower in this field – hence requiring some problems to be overcome by trial and error in the first instance.

The obvious question raised was "could this enormous expenditure in R & D be justified?" A negative retort is apparent when present SCP production output is quantified. However, qualitative considerations of consequential influences to fermentation technology, development and methodology yield a less negative answer (see Table 33).

The challenge of achieving large-scale, continuous, aseptic, and aerobic operation demanded the development of expertise in numerous fields – mechanical, chemical and biochemical engineering – to facilitate equipment design, process integration and optimization of unit operations, electronic and control engineering to provide optimal automation and reliable sensoring techniques. In addition, the development of the biological sciences in microbiological techniques, genetic manipulation, strain stabilization and the like is required.

Future trends in SCP production are discussed from three viewpoints, namely, 1) view of the function of SCP as a protein source, 2) the preferred raw material or substrate, and 3) the type of SCP technology to be used for future commercialization.

Single cell protein may be used as either a food/feed or a food/feed additive. Industrial uses of any protein, especially in the food industry, is based on the ability of the protein to foam, to emulsify or to gel. Frequently, the food industry utilizes egg proteins for these purposes, although vegetable and milk proteins are used increasingly as substitutes [17]. Single cell protein has the required functional

Table 33. Contributions of SCP research and development programs to the overall development of fermentation technology. (Based in part on Hamer and Harrison [38] and on Hamer [615])

- The demonstration of large-scale, stable chemostatic, continous flow aerobic cultures
- The challenge of aseptic processing on high tonnage scale
- Control of composition of product by use of precise feeding techniques
- The introduction of various column fermentor designs in order to improve mass transfer
- Enhanced understanding of physical environmental influences on fermentative growth and production processes
- The evaluation of traditional cell/liquid separation techniques and their subsequent replacement by new methods
- The demonstration of the operation of high productivity biological systems for long periods
- The introduction of process water recycle to biological industries
- The widespread acceptance of the potential of non-carbohydrate carbon and energy substrates for fermentation processes
- The re-establishment of fermentation processes for bulk chemical production
- The introduction of defined mixed cultures for product synthesis, offering a viable alternative to monocultures
- The first commercial application of recombinant DNA techniques
- The implementation of microprocessors for controlling sterilization and fermentation processes
- Enhanced understanding of fermentation kinetics
- The widening of knowledge of physiology and biochemistry of microorganisms utilizing a variety of carbon substrates
- Enhanced knowledge in flavor, texture and digestive properties of food

properties and could be used as a substitute if the respective downstream technologies were developed.

Presently, the amount of SCP used as a human food is limited. Micro-algae pressed into tablet forms are sold as a health food product in the Far East, and to some extent in the USA. "Mycoprotein", manufactured in England by Ranks Hovis McDougall Limited (RHM) (Table 24) has U.K. approval as a human food. Unlike many other SCP's, it is a low cost, texturized material lacking the off-flavors of vegetable proteins and having excellent nutritional qualities. Production of "Mycoprotein" may be successful economically if accepted as a meat substitute, as it could then be sold at a high price, similar to that of the algal products sold as health food. However, to achieve commercialization of "Mycoprotein", RHM had to invest considerable time (15 years) and money [8]. Furthermore, until effects of "Mycoprotein" on the market have been evaluated, the production scale would be limited.

The constraints discussed above suggest that SCP is not likely to be widely used as a human food in the next few years, and thus the short term aim is to capture part of the animal feed market in replacing fish and soy meal as feed additives. If SCP can compete with and sell alongside soybean and fish meal, their concentrates, and skimmed milk powder, then this will demand very large production volumes of SCP. Unfortunately, current conditions and the present state of the art show that SCP cost cannot compete with the prices of conventional protein sources, especially that of soybean meal. It can be said that almost no SCP production plant is running economically at present. The Western free market proves this point by forcing SCP plants off the market except where noncommercial priorities dominate (see Tables 3–5). Beyond the Iron Curtain, the Soviet Union and other Eastern European countries has developed massive SCP production plants. This phenomenon is due to the communistic economy, shortage of foreign exchange, agricultural limitations and a very high protein demand [8].

Future trends in substrate and technology choice for SCP production are interlinked, such that the substrate type defines the processing technology requirements. Processes based on the currently expensive oil-based products were developed in the last three decades, and a large scale methanol-based process is operational (Table 24). It was thought originally that oil-based substrates would be indefinitely available in large quantities and at low prices. The 1973 oil crisis proved these assumptions to be invalid. Since the substrate costs comprise a large part of the plant operating costs (Table 27), most petroleum-based processes became uneconomical, even when sophisticated technology was used to improve process economy by having a highly productive, aseptic, continuous flow, large-scale (> 50,000 ton/yr) system with water recycle, low effluents and high-yielding microorganism on the given substrate, as was previously discussed [269]. Other disadvantages of using petroleum-based substrates are the long-term political instability of the Persian Gulf countries (leading to the bitter and lasting war between Iran and Iraq), and the fluctuating policies of Western countries. The continued trend, therefore, must be that we look for cheaper carbon sources whose availability does not depend on political outlook. These are likely to be plant carbohydrates (renewable resources).

Considerations show that production of SCP for animal feeds from nonpolymerized carbohydrates is not economical. In 1981, soy meal, consisting of 45% "true protein", cost $ 205/ton and sugar cost about $ 200/ton [8, 60, 804]. One ton of protein obtained from about 1.6 tons of microbial biomass requires 3.3 tons of sugar substrate, assuming a yield of 0.5 g cell dry weight/g sugar utilized. This would make the basic cost of single cell protein $ 660/ton, compared to $ 450/ton of soy protein.

Another option is the use of cheap carbohydrate industrial "waste" materials which cost about 10% of the nonpolymerized sugar price. These substrates suffer from intermittent availability and slow rates of conversion, particularly in the case of lignocellulose. Industry is finding more uses for such byproducts and wastes (Fig. 32) and their market is consequently expanding.

Two main types of technologies are practical for use of lignocellulosic wastes for SCP production, namely, liquid and solid phase fermentations. Liquid phase fermentations are aerobic and use hydrolyzates containing both hexoses and pentoses. Ideally, these processes can be operated on a small scale ("village type" process) in rural area (see the "Waterloo" process described in Table 24), or on a large scale to treat effluents from large agroindustrial plants (exemplified by the ICAITI process, intended to produce fungal-SCP from wastes obtained in coffee "beneficios", see Table 5). These technologies have low capital inputs but high manpower inputs and, therefore, are suited to the developing countries. Ideally, the domestic wastes produced in abundance in a certain country would be utilized there.

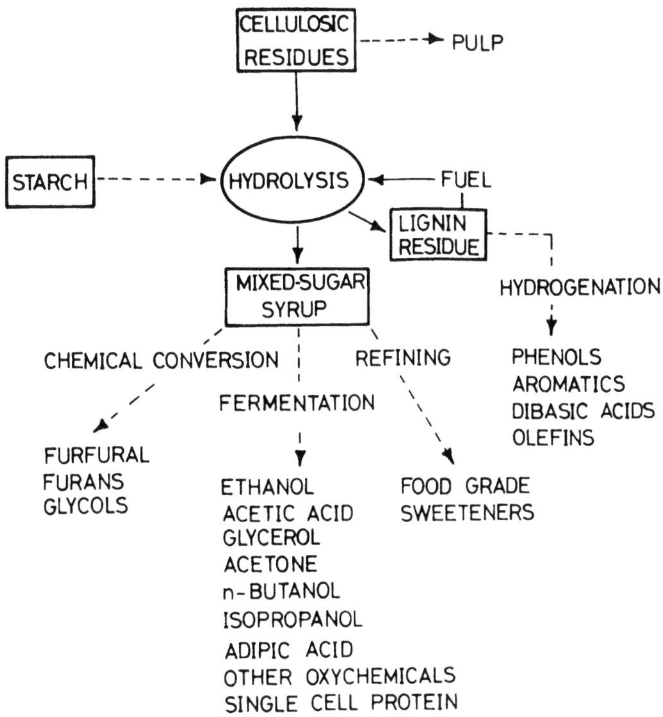

Fig. 32. Feedstocks from renewable resources. (Courtesy of Dr. Busche)

Solid and semisolid fermentations of lignocellulosic wastes may be aerobic or anaerobic. Their function is to reduce fiber content yet increase protein content of the waste so that it may be used directly as an animal feed (exemplified in Table 6 and discussed further by Farranti and Fiechter [28]). Such processes should be operated in rural areas of developing countries though the process should have proven successful in upgrading the protein content of the product.

Another view of SCP production is that it should be one of a multifaceted production system, it being a byproduct [8, 16, 17]. One multiproduct system is the production of antibiotics, bio-polymers, steroids, organic acids, and amino acids. Another is the anaerobic fermentation of carbohydrates for production of methane, ethanol and organic acids. A multipurpose system is exemplified also by the reduction of the biological oxygen demand from industrial or food processing or agricultural wastes.

A large tonnage (270,000 tons) of microbial biomass can be produced per 100 million gallons gasohol from a corn feedstock (according to Solomons [8]). In Brazil, for example, there are massive intensive programs for ethanol production from sugar cane and cassava, and in the USA, the energy policy under the Carter Administration established a 1990 goal of 6 billion gallon/year of ethanol from corn for use in gasohol [413]. Thus, worldwide, the potential quantities of microbial biomass produced as byproducts will be very large and should be sold at cheaper rates than those of SCP's produced as the major product. It should be noted that the nutritive quality of a "byproduct SCP" is a matter of chance and does not compare with the consistently high standard of "mainproduct SCP". The yield and productivities in "byproduct SCP" processes are not maximized and are sometimes intentionally limited such that the main product yield is not reduced. These "byproduct processes" do not apply the same scientific principles used in the "mainproduct processes" and consequently did not merit much discussion in this monograph.

Recently, Batt and Sinskey [41] introduced a very intriguing idea, based on the finding that a temperature-sensitive *S. cerevisiae* mutant which could rapidly effect autolysis could be isolated through chemical mutagenesis [805]. Batt and Sinskey [41] suggested that the introduction of a gene coding for a lytic enzyme, under automatically triggered yeast promotor control, could allow formation of an autolytic strain of *S. cerevisiae*. Moreover, this external triggering of a gene function would facilitate the design of two-stage processes where the main metabolite, such as ethanol, would be extracted from a first stage, and the spent yeast would then be utilized for SCP production in a secondary fermentation with appropriately timed triggering. The triggering agent (whether physical or chemical) would activate the gene(s) coding for cell lysis, morphological changes which cause faster cell precipitation, or cause overproduction of a protein enriched in a specific amino acid. The process could thus be "tailored" to matrix specifications, though such a process would have to be proven successful in practice.

Since future SCP production trends are based on future supply and demand of raw materials, government policies, import-export trends, and new scientific achievements, the reliability of predictions is partially accurate at best. For example, soy beans are produced mainly in the USA, and partly in Brazil. Most countries depend on the producers' exports. The growing dependence of developing countries on grain imports has been discussed [9], e. g., between the years 1966–1970 and 1976–1980, food imports to these countries increased at an average annual rate of 6.4%. More than 16 million ton/yr of feed protein are imported into Europe, mostly from the USA (soybean meal) and South America (soybean meal from Brazil and fishmeal from Peru). France, for example, consumes 1.7 million tons of protein annually and produces only 30% of this amount. It is necessary, therefore, to import about 2.5 million tons of soya cakes annually [803].

Western Europe produces only 0.5% of its protein requirements in the form of SCP, thus relying to a great extent both financially and politically on the USA for the import of conventional proteins. This situation makes any alternative to the costly protein imports attractive [806]. The production of SCP may also have a stabilizing effect on the fluctuating costs of oilseed and fish meals which would greatly benefit those countries that depend on protein imports [722]. The aforementioned factors should encourage, and in certain situations, may force various countries to develop comprehensive strategic policies for protein production (not necessarily SCP) in order to promote self-sufficiency and reduce political and economic pressures.

In terms of political economics, the dilemma facing Israel, for example, can be approached from several directions. Israel imports practically all its animal protein feeds (about one million ton/yr of soybean meal), and is consequently particularly sensitive both to political implications, especially since it is a small country, and to the unstable costs of animal protein feeds on the international market. Recognizing these realities and trying to reduce political dependency, the government could decide to consider the increase of locally produced protein as a political priority. Accordingly, the government could decide to allocate land for the cultivation of soybean, instead of cultivating the present major crop of cotton or, based on an inhouse technology (Table 4) and acquired knowledge, could produce single cell protein on a large commercial scale.

This political decision might be taken even if the SCP production plant(s) would be uneconomical and even if the government would lose money by not exporting cotton which is a higher added value crop than soybean. However, it should be noted that in this situation, the SCP product would have to compete with locally grown soybean, and not with an imported protein. The price of the locally-grown soybean would probably be much higher than the imported soybean (due to water and arable land shortage and high energy costs in Israel); thus the local SCP process could utilize a high cost raw material and technology and still be economical, relative to the soybean alternative.

During the last years, it became apparent that the protein shortage became a fact of life in many developing countries; yet the major incentive for SCP production is an economic one and is pronounced in Europe, the USSR, Japan, and the Middle East. This is explained by the increasing demand for protein-rich raw materials for incorporation into compounded animal feeds that exists in these countries and by their ability to maintain greater economic stability through assured sources of energy [38, 626]. This issue was discussed in detail by Hamer and Harrison [38] for the Middle East Arab countries, net exporters of hydrocarbons and one of the net importers of food and feed.

Recently, the Federation of the Arab Scientific Research Councils issued a statement declaring that the Arab countries, being fully aware of their nutrition-dependence problem, intendend to solve it by developing large-scale SCP production plants with fossil-based (hydrocarbons) raw materials as carbon sources [807], see also [695] and [808] (the first Arab Gulf conference on biotechnology and applied microbiology with emphasis on SCP, was held at Riyadh, Saudi Arabia, in November 1984).

It was stated that a major protein crisis can arise if principles of Western economics dominate the decision-making. These principles dictate only a limited increase in SCP production, with the worldwide protein demand still escalating [38]. However, this pessimistic view is not the conclusion to be drawn here. It is our belief that the future capabilities in agriculture and even more in biotechnology will not be totally limited by economic considerations.

The world coped with the 1973 oil crisis – "energy at high cost" – and should likewise overcome monetary influences and press for "food at all costs". The consequential surge of new SCP technologies should then provide one or two processes which are economical or nearly so, keeping in mind that cost estimates for SCP production are known to have a notoriously short half-life, and should therefore not be treated as gospel.

Changes in the world's economic climate, together with current research and development efforts, will lead to significant improvements in the economic viability of SCP processes. It is this which strengthens the belief of both scientists and laymen that, in the future, microbial biomass will make a more valuable contribution to animal and human diets, and will solve, at least in part, the world's acute protein shortage.

5 References

1. Mateles RI, Tannenbaum SR (eds) (1968) Single cell protein I. MIT Press, Cambridge, MA
2. Altschul AM (1968) The agricultural, scientific, and economic basis for low-cost protein food. In: Mateles RI, Tannenbaum SR (eds) Single cell protein I. MIT Press, Cambridge, MA, p 48
3. Brown LR (1968) World food problems. In: Mateles RI, Tannenbaum SR (eds) Single cell protein I. MIT Press, Cambridge, MA, p 11
4. Munro HN (1968) The nature of protein needs. In: Mateles RI, Tannenbaum SR (eds) Single cell protein I. MIT Press, Cambridge, MA, p 27
5. Scrimshaw NS (1975) Single-cell protein for human consumption – an overview. In: Tannenbaum SR, Wang DIC (eds) Single cell protein II. MIT Press, Cambridge, MA, p 24
6. Borgstrom G (1976[2]) Principles of food science, vol 2, food microbiology and biochemistry. Food and Nutrition Press, Inc., Westport, CT
7. Litchfield JH (1983) Science 219:740
8. Solomons GL (1983) CRC Crit Rev Biotech 1:21
9. Mellor JW, Adams RH Jr (1984) Chem Eng News 62:32
10. Scrimshaw NS (1968) Introduction. In: Mateles RI, Tannenbaum SR (eds) Single cell protein I. MIT Press, Cambridge, MA, p 3
11. Litchfield JH (1978) Chemtech 8:218
12. Rose AH (1979) History and scientific basis of large-scale production of microbial biomass. In: Rose AH (ed) Economic microbiology, vol 4. Microbial biomass. Academic Press, London, p 1
13. Peppler HJ (1979[2]) Production of yeasts and yeast products. In: Peppler HJ, Perlman D (eds) Microbial technology, vol 1. Microbial processes. Academic Press, New York, p 157
14. Rose AH (ed) (1979) Economic microbiology, vol 4. Microbial biomass. Academic Press, London
15. Lipinsky ES (1981) Science 212:1465
16. Rolz C (1983) IOBB Newsletter, Sept Issue, p 1
17. Jarvis B, Holms AW (1982) J Chem Tech Biotechnol 32:224
18. Anthony C (1982) The biochemistry of methylotrophs. Academic Press, London
19. Pontanel HG De (ed) (1972) Proteins from hydrocarbons. Academic Press, New York
20. Gutch S (1973) Proteins from hydrocarbons. Noyes Data Corporation, Park Ridge, NJ
21. Davis JP (ed) (1974) Single cell protein. Academic Press, New York
22. Tannenbaum SR, Wang DIC (eds) (1975) Single cell protein II, MIT Press, Cambridge, MA
23. Wagner F (ed) (1975) Symposium mikrobielle Proteingewinnung. Verlag-Chemie, Weinheim
24. Rockwell PJ (1976) Single cell proteins from cellulose and hydrocarbons, Noyes Data Corporation, Park Ridge, NJ
25. Gaden EL Jr, Humphrey AE (eds) (1977) Single cell protein from renewable and nonrenewable resources. Biotech Bioeng, Symp 7. John Wiley & Sons, New York
26. Shelef G, Soeder CJ (eds) (1980) Algae biomass production and use. Elsevier/North-Holland, New York
27. Harrison DEF, Higgins IJ, Watkinson R (eds) (1980) Hydrocarbons in biotechnology. Heyden & Sons Ltd, London
28. Ferranti MP, Fiechter A (eds) (1983) Production and feeding of single cell protein. Applied Science Publishers Ltd, London
29. Lipinsky ES, Litchfield JH (1970) Crit Rev Food Technol 1:581
30. Snyder HE (1970) Adv Food Res 18:85
31. Kihlberg R (1972) Ann Rev Microbiol 26:427

32. Waslien CI (1975) Crit Rev Food Sci Nutr 6:77
33. Litchfield JH (1977) Adv Appl Microbiol 22:267
34. Riviere J (1977) Industrial applications of microbiology. John Wiley & Sons, New York
35. Cooney CL, Rha C, Tannenbaum SR (1979) Adv Food Res 26:1
36. Mateles RI (1979) Symp Soc Gen Microbiol 29:29
37. Hamer G, Hamdan IY (1979) Chem Soc Rev 8:143
38. Hamer G, Harrison DEF (1980) Single cell protein: the technology, economics and future potential. In: Harrison DEF, Higgins IJ, Watkinson R (eds) Hydrocarbons in biotechnology. Heyden & Sons Ltd, London
39. Reed G (1982) Microbial biomass, single cell protein, and other microbial products. In: Reed G (ed) Prescott and Dunn's industrial microbiology. Avi Publishing Comp Inc, Westport, CT, p 541
40. Samuelov NS (1983) Adv Biotech Processes 1:293
41. Batt CA, Sinskey AJ (1984) Food Technol 38:108
42. Delbrück M (1910) Wochschrift für Brauerei 27:375
43. Hockberg M, Melnick D, Oser BL (1945) J Nutr 26:385
44. Litchfield JH (1968) The production of fungi. In: Mateles RI, Tannenbaum SR (eds) Single cell protein I. MIT Press, Cambridge, MA, p 309
45. Anderson C, Longton J, Maddix C, Scammel GW, Solomons GL (1975) The growth of microfungi on carbohydrates. In: Tannenbaum SR, Wang DIC (eds) Single cell protein II. MIT Press, Cambridge, MA, p 314
46. Benemann JR, Weissman JC, Oswald WJ (1979) Algal biomass. In: Rose AH (ed) Economic microbiology, vol 4. Microbial biomass. Academic Press, London, p 177
47. Jüttner F (1982) Process Biochem 17:2
48. Soong P (1980) In: Shelef G, Soeder CJ (eds) Algae biomass production and use. Elsevier/North-Holland, New York, p 92
49. Lipinsky ES, Litchfield JH (1974) Food Technol 28:16
50. Clement G (1975) Producing *Spirulina* with CO_2. In: Tannenbaum SR, Wang DIC (eds) Single cell protein II. MIT Press, Cambridge, MA, p 467
51. Laskin AI (1977) Biotech Bioeng Symp 7:91
52. Vonshak A, Boussiba S, Abeliovich A, Richmond A (1983) Biotech Bioeng 25:341
53. Azov Y, Shelef G, Maraine R (1982) Biotech Bioeng 24:579
54. Ben-Amotz A, Avron M (1983) Ann Rev Microbiol 37:95
55. Gow JS, Littlehailes JD, Smith SRL, Walter RB (1976) SCP production from methanol: bacteria. In: Tannenbaum SR, Wang DIC (eds) Single cell protein II. MIT Press, Cambridge, MA, p 370
56. Faust U, Prave P, Sukatsch DA (1977) J Ferment Technol 55:609
57. Mogren H (1979) Process Biochem 14:2
58. Ebbinghaus L, Ericsson M, Lindblom M (1981) Production of single cell protein from methanol by bacteria. In: Moo-Young M (ed) Advances in biotechnology II. Pergamon Press, Toronto, p 413
59. Ridgeway JA Jr, Lappin TA, Benjamin BM, Corns JB, Akin C (1975) US Patent 3.865.691
60. Humphrey AE (1981) Production of single cell protein from ethanol. In: Moo-Young M (ed) Advances in biotechnology II. Pergamon Press, Toronto, p 431
61. Ko PC, Yu Y (1968) Production of SCP from hydrocarbons: Taiwan. In: Mateles RI, Tannenbaum SR (eds) Single cell protein I. MIT Press, Cambridge, MA, p 255
62. Dube P, Gupta A, Khan WA, Dutta KK, Kaw JL, Dikshith TSS, Mehrotra NK, Pandya KP (1982) XIII International congress in microbiology, Boston, MA, p 151
63. Watts H (1976) Chemistry and Industry 3:537
64. LeDuy A (1976) Process Biochem 14:5
65. Knecht R, Prave P, Seipenbusch R, Sukatsch DA (1977) Process Biochem 12:11
66. Birckenstaedt JW, Faust U, Sambeth W (1977) Process Biochem 12:7
67. Shacklady CA (1970) Outlook on Agriculture 6:102
68. O'Sullivan DA (1978) Chem Eng News 3:12
69. Cooper P, Silver R, Boyle J (1975) Semi-commercial studies of a petroprotein process based on n-paraffins. In: Tannenbaum SR, Wang DIC (eds) Single cell protein II. MIT Press, Cambridge, MA, p 454
70. Zanetti R (1984) Chem Eng 91:18

71. Forage AJ (1978) Process Biochem 13:8
72. Davy CAE, Wilson D, Lyon JCM (1981) Commercial production of feed yeast from carbohydrate waste. In: Moo-Young M (ed) Advances in biotechnology II. Pergamon Press, Toronto, p 343
73. Jarl K (1969) Food Technol 23:23
74. Lemmel SA, Heimsch RC, Edwards LL (1979) Appl Environ Microbiol 37:227
75. Sanchez-Marroquin A (1977) Biotech Bioeng Symp 7:23
76. Espinosa R (1976) Chemtech 6:636
77. Romantschuk H, Lehtomaki M (1978) Process Biochem 13:16
78. Rychtera M, Barta J, Fiechter A, Einsele AA (1977) Process Biochem 12:26
79. Anderson R, Weisbaum RB, Robe K (1974) Food Process 35:58
80. Bernstein S, Tzeng CH, Sisson D (1977) Biotech Bioeng Symp 7:1
81. Kim JH, Lebault JM (1981) Eur J Appl Microbiol Biotechnol 13:151
82. Bayer K, Meyrath J (1976) Zentralblatt Bakteriol Hygiene 252:37
83. Moulin G, Malige B, Galzy P (1983) Industrial production of SCP from whey. In: Ferranti MP, Fiechter A (eds) Production and feeding of single cell protein. Applied Science Publishers, London, p 153
84. Robe K (1964) Food Process 25:95
85. Kappeli O, Halter N, Puhan Z (1981) Upgrading of milk ultrafiltration permeate by yeast fermentation. In: Moo-Young M (ed) Advances in biotechnology II. Pergamon Press, Toronto, p 351
86. Halter N, Puhan Z, Kappeli O (1983) Upgrading of milk UF-permeate by yeast fermentation-semiindustrial trials and economy. In: Ferranti MP, Fiechter A (eds) Production and feeding of single cell protein. Applied Science Publishers, London, p 147
87. Chahal DS, Vlach D, Moo-Young M (1981) Upgrading the protein feed value of lignocellulosic materials using *Chaetomium cellulolyticum* in solid-state fermentation. In: Moo-Young M (ed) Advances in biotechnology II. Pergamon Press, Toronto, p 327
88. Dimmling W (1975) Hydrocarbon Process 9:169
89. Glanser M, Ban SN (1984) Process Biochem 19:105
90. Zomer E, Klein D, Rozhansky M, Er-el Z, Joson LM, Goldberg I (1981) Biotech Lett 3:513
91. Kamibubo T, Tanaka M, Taniguchi M, Morita T, Matsuno R (1981) Production of SCP from waste cellulose. In: Moo-Young M (ed) Advances in biotechnology II. Pergamon Press, Toronto, p 311
92. Kositanont C, Charoensiri K, Bhumiratana A (1981) Single cell production from cellulosic material. In: Taguchi H (ed) Microbial utilization of renewable resources II. Osaka University, Osaka, p 47
93. Davey G, Bruce J (1982) Biotech Bioeng 25:647
94. Tuse D, Russell LA, Hsieh DPH (1981) Nutritional and toxological evaluation of SCP produced from an environmental waste. In: Moo-Young M (ed) Advances in biotechnology II. Pergamon Press, Toronto, p 363
95. Rao M, Mishra C, Seeta R, Srinivasan MC, Deshpande VV (1983) Biotech Lett 5:301
96. Araujo A, D'Souza J (1980) J Ferment Technol 58:399
97. El-Masry HG (1983) J Sci Food Agric 34:725
98. Kim BH, Wimpenny JWT (1981) J Ferment Technol 59:275
99. Ek M, Eriksson KE (1980) Biotech Bioeng 22:2273
100. Noparatnaraporn M, Nishizawa Y, Nagai S (1980) Utilization of photosynthetic bacteria on soybean waste. In: Taguchi H (ed) Annual reports of international center of cooperative research and development in microbial engineering. Osaka University, Osaka, p 238
101. Horitsu H, Hashimoto T, Tokunaka T, Kawai K, Wakazono H (1982) XIII International congress of microbiology. Boston, MA, p 151
102. Garg SK, Neelakantan S (1982) Biotech Bioeng 24:2407
103. Enriquez A, Rodrigues H (1983) Biotech Bioeng 25:877
104. Sidhu MS, Sandhu DK (1980) Biotech Bioeng 22:689
105. Bottaro Castela R, Waehner RS, Giulietti AM (1984) Biotech Lett 6:195
106. Moresi M, Marchionni G (1983) Biotech 1073
107. Kargi F, Schuler ML (1981) Biotech Lett 3:409
108. Kargi F, Schuler ML, Vashon R, Seeley HW, Henry A, Austic RE (1980) Biotech Bioeng 22:1567
109. Oron G, Shelef G, Levi A (1979) Biotech Bioeng 21:2169

110. Margaritis A, Bajpai P, Cannell E (1981) Biotech Lett 3:595
111. Apaire V, Guiraud JP, Galzy P (1983) Z Allg Mikrobiol 23:211
112. Illanes A, Schaffeld G (1983) Biotech Lett 5:305
113. Friedrich J, Cimerman A, Perdih A (1983) Eur J Appl Microbiol Biotechnol 17:243
114. Barker TW, Patton AM, Marchiant R (1983) J Sci Food Agric 34:638
115. Sethi RP, Grainger JM (1981) Conversion of mangostone into SCP for animal feed by *Aspergillus niger* using solid substrate fermentation. In: Moo-Young M (ed) Advances in biotechnology II. Pergamon Press, Toronto, p 319
116. Quinn JP, Marchant R (1979) Eur J Appl Microbiol Biotechnol 6:251
117. DeLeon R, Calzada F, Herrera R, Rolz C (1980) J Ferment Technol 58:579
118. Chung SL, Meyers SP (1979) Dev Ind Microbiol 20:723
119. Azoulay E, Jouanneau F, Bertrand JC, Raphael A, Janssens J, Lebeault JM (1980) Appl Environ Microbiol 39:41
120. Azoulay E, Janssens J, Raphael A, Bertrand JC (1981) Growth of *Candida tropicalis* on starch and cassava powder for the obtention of high-protein feed. In: Moo-Young M (ed) Advances in biotechnology II. Pergamon Press, Toronto, p 333
121. Noparatnaraporn M, Sasaki K, Nishizawa Y, Hayashi M, Nagai S (1982) XIII International congress of microbiology. Boston, MA, p 150
122. Wilson JJ, Khachatourians GG, Ingledew WM (1982) Biotech Lett 4:333
123. Spencer-Martins I, van Uden N (1979) Eur J Appl Microbiol Biotechnol 6:241
124. Levy-Rick S, Champagne CC, Yaguchi M, Lusena CV, Calleja GB, Nasim A (1982) Can Soc Microbiol 116
125. Nga BH (1982) Improvement of utilization of starch by strains of *Aspergillus niger*. In: Taguchi H (ed) Annual reports of international center of cooperative research and development in microbial engineering. Osaka University, Osaka, p 455
126. Stakheyev IV, Babitskaya VG, Kostina AM, Vadetskii BY (1983) Mikol Fitopatol 17:135
127. Moresi M, Solinas MA, Matteucci S (1983) Eur J Appl Microbiol Biotechnol 18:92
128. Pujol F, Bahar S (1983) Eur J Appl Microbiol Biotechnol 18:361
129. Urakami T, Terao I, Nagai I (1983) J Ferment Technol 61:221
130. Miura Y, Okazaki M, Komemushi S, Sakata T, Shiroza S, Obana S (1983) U.K. Patent Appl G.B. 2120274
131. Cabane B, Galzy P (1983) US Patent 4416987
132. Couderc R, Baratti J (1980) Agr Biol Chem 44:2279
133. Allais JJ, Louktibi A, Baratti J (1983) Agr Biol Chem 47:1509
134. Hamer G, Hamdan IY, Humphrey AW (1981) A route for SCP production using volatile liquid hydrocarbon feedstocks. In: Moo-Young M (ed) Advances in biotechnology II. Pergamon Press, Toronto, p 395
135. Katrush RV, Kozlova LI, Razhkova MI, Zhdannikova EN, Velikoslavinskaya OI, Bauch J, Gentzsch H, Bohlmann I (1981) Production of feed yeast from petroleum distillates. In: Moo-Young M (ed) Advances in biotechnology II. Pergamon Press, Toronto, p 401
136. Sprenger B, Rehm HJ (1983) Eur J Appl Microbiol Biotechnol 17:265
137. Zarilla KA, Perry JJ (1984) Arch Microbiol 137:286
138. Wlodarczyk Z (1981) Chemical composition of yeasts cultivated in various hydrocarbon substrates. In: Moo-Young M (ed) Advances in biotechnology II. Pergamon Press, Toronto, p 447
139. Krasnikov EI, Rudaya SP, Nagornaya SS, Klyushnikova TM, Ivanova LL, Sumnevich VG (1983) Mikrobiologiya 45:24
140. Kilian SG, Prior BA, Lategan PM, Kruger WCJ (1981) Biotech Bioeng 23:267
141. Kamel BS (1979) Process Biochem 14:12
142. Braun R, Meyrath J, Stuparek W, Zerlauth G (1979) Process Biochem 14:16
143. Prior BA (1984) Biotech Bioeng 26:748
144. Prior BA, Botha M, Custers M, Casaleggio C (1981) Fermentation of pineapple effluent by *Candida utilis*. In: Moo-Young M (ed) Advances in biotechnology II. Pergamon Press, Toronto, p 337
145. Rale VB (1984) Eur J Appl Microbiol Biotechnol 19:106
146. Revah-Moiseev S, Carroad PA (1981) Biotech Bioeng 23:1067
147. Hang YD (1980) Appl Environ Microbiol 39:470
148. Sandhu DK, Waraich MK (1983) Biotech Bioeng 25:797

149. Foda MS (1983) Egypt J Microbiol 18:151
150. Moresi M, Colicchio A, Sansovini F (1980) Eur J Appl Microbiol Biotechnol 9:173
151. Moresi M, Colicchio A, Sansovini F, Sebastiani E (1980) Eur J Appl Microbiol Biotechnol 9:261
152. Castillo FJ, Sanchez S, Sanchez L, Bales S, Goncalves JA (1981) Utilization of milk whey. In: Moo-Young M (ed) Advances in biotechnology II. Pergamon Press, Toronto, p 686
153. Abou-Zeid AZA, Abd El-Fattah AF, Farid MA (1979) Agr Biol Chem 43:1977
154. Gold D, Mohagheghi A, Cooney CL, Wang DIC (1981) Biotech Bioeng 23:2105
155. Noonai A, Flegel TW (1981) Biomass production from acid hydrolysate of spent sulfite liquor. In: Taguchi H (ed) Microbial utilization of renewable resources II. Osaka University, Osaka, p 33
156. Oliva RU, Hang YD (1979) Appl Environ Microbiol 38:1027
157. Henry DP, Greenfield PF, Thomson RH (1983) Eur J Appl Microbiol Biotechnol 18:109
158. Welsh FW, Zall RR (1984) Process Biochem 19:122
159. Ercoli E, Ertola R (1983) Biotech Lett 5:457
160. Martinet F, Ratomahenina R, Graille J, Galzy P (1982) Biotech Lett 4:9
161. Koh JS, Kodama T, Minoda Y (1983) Agr Biol Chem 47:1207
162. Koh JS, Kodama T, Yamakawa T, Minoda Y (1980) Microbial utilization of palm oil – screening of yeasts and cultural conditions of jar fermentor for cell production. In: Taguchi H (ed) Annual reports of international center of cooperative research and development in microbial engineering. Osaka University, Osaka, p 167
163. Nakahara T, Sasaki K, Tabuchi T (1982) J Ferment Technol 60:89
164. Montet D, Ratomahenina R, Ba A, Pina M, Graille J, Galzy P (1983) J Ferment Technol 61:417
165. Karthigesan J, Brown BS (1981) J Chem Technol Biotechnol 31:55
166. Vrati S (1984) Eur J Appl Microbiol Biotechnol 19:199
167. Pirt SJ, Lee YK, Richmond A, Pirt MW (1980) J Chem Technol Biotechnol 30:25
168. Laws EA, Terry KL, Wickman J, Chalup MS (1983) Biotech Bioeng 25:2319
169. Miura Y, Okazaki M, Ohi K, Nishimura T, Komemushi S (1982) Biotech Bioeng 24:1173
170. Siegel RS, Ollis DF (1984) Biotech Bioeng 26:764
171. Stanier RY, Doudoroff M, Adelberg EA (1970[3]) The microbial world. Prentice Hall, Englewood Cliffs, NJ
172. Shipman RH, Fan LT, Kao IC (1977) Adv Appl Microbiol 21:161
173. Mahler HR, Cordes EH (1967[3]) Biological chemistry. Harper and Row, Evanston and London, New York
174. Kobayashi M, Kurata SI (1981) Process Biochem 16:27
175. Becker EW (1981) Process Biochem 16:10–14
176. Ludwig HF, Oswald WJ (1952) Scientific Monthly 74:3
177. Oswald WJ, Golueke CG (1968) Large-scale production of algae. In: Mateles RI, Tannenbaum SR (eds) Single cell protein I. MIT Press, Cambridge, MA, p 271
178. Soeder CJ, Binsack R (1978) Ergebn Limnol 11
179. Nigam BP, Ramanathan PK, Venkataraman LV (1981) Biotech Lett 3:619
180. Eisenberg DM, Koopman B, Benemann JR, Oswald WJ (1981) Biotech Bioeng Symp 11:429
181. Aiba S, Ogata T (1977) J Gen Microbiol 102:179
182. Pirt SJ (1981) Fermentation – a question of life. In: Moo-Young M (ed) Advances in biotechnology II. Pergamon Press, Toronto, p XVII
183. Repaske R (1962) J Bacteriol 83:418
184. Repaske R (1966) Biotech Bioeng 8:217
185. Lafferty RM, Schlegel HG (1971) Adv Biochem Eng 1:143
186. Goto E, Suzuki K, Kodama T, Minoda Y (1977) Agr Biol Chem 41:521
187. Goto E, Kodama T, Minoda Y (1971) Agr Biol Chem 41:685
188. Morinaga Y, Yamanaka S, Ishizaki A, Hirose Y (1978) Agr Biol Chem 42:439
189. Miura Y, Okazaki M, Ohi K, Nishimura T (1981) Agr Biol Chem 45:1181
190. Lafferty RM, Moser A, Steiner W (1975) Production of single cell protein from chemo-lithotropic bacteria. In: Wagner F (ed) Microbial production of protein. Verlag-Chemie, Weinheim, p X
191. Cangelosi GA, Wheelis ML (1984) J Bacteriol 159:138
192. Schobert P, Bowien B (1984) J Bacteriol 159:167

193. Schuster E, Schlegel HG (1967) Arch Microbiol 58:380
194. Robra KH, Lafferty RM, Schlegel HG (1972) Zbl Bakt II Abtlg 127:651
195. Miyoshi M (1985) J Wiss Bot 28:269
196. Söhngen NL (1906) Zentr Bakt Parasitenk 15:513
197. Söhngen NL (1913) Zentr Bakt Parasitenk 37:595
198. Tausson WO (1928) Neftyanoe Khoz 14:220
199. Mogilevskii GA (1940) Rozvedka Okhrana NEDR 12:30
200. Hassler GL (1943) U.S. Patent 2.321.292
201. Humphrey AW (1967) Biotech Bioeng 9:3
202. Fuhs GW (1961) Arch Microbiol 39:374
203. Bos P, de Bruyn JC (1973) Antonie van Leevwenhoek 39:99
204. Walker JD, Austin HF, Colwell RR (1975) J Gen Appl Microbiol 21:27
205. Levi JD, Shennan JL, Ebbon GP (1979) Biomass from liquid n-alkanes. In: Rose AH (ed) Economic microbiology, vol 4. Microbial biomass. Academic Press, London, p 362
206. Atlas RM (1981) Microbiol Rev 45:180
207. Komagata K, Kakase T, Katsuya N (1964) J Gen Appl Microbiol 10:313
208. Markovetz AJ, Kallio AE (1964) J Bacteriol 87:968
209. Scheda R, Bos P (1966) Nature 211:660
210. Klug MJ, Markovetz AJ (1971) Utilization of aliphatic hydrocarbons by microorganisms. In: Rose AH, Wilkinson JF (eds) Advances in microbial physiology. Academic Press, Inc, New York, p 1
211. Casas-Campillo C, Morales JL, Larrea S (1971) Single-cell protein from filamentous fungi propagated on hydrocarbons. In: Freitas YM, Fernandes F (eds) Global impacts of applied microbiology. University of Bombay, Bombay, p 531
212. Lowery CE Jr, Foster JW, Jurtshuk P (1968) Arch Microbiol 60:246
213. Markovetz AJ Jr, Cazin J, Allen JE (1968) Appl Micriobiol 16:487
214. Champagnat A, Vernet C, Laine B, Filosa J (1963) Nature (London) 197:13
215. Done K (1978) Financial Times (London), July 20th
216. Litchfield JH (1977) Biotech Bioeng Symp 7:77
217. Dimmling W, Spiepenbusch R (1978) Process Biochem 13:9
218. Johnson MJ (1964) Chem Indus 9:1532
219. Fencl Z (1969) Biotech Bioeng Symp 1:63
220. Munk V, Dostalek M, Volfova O (1969) Biotech Bioeng 11:383
221. Anon (1967) Chem Eng News 9:46
222. Ko PC, Yu Y (1968) Production of SCP from hydrocarbons. In: Mateles RI, Tannenbaum SR (eds) Single cell protein I. MIT Press, Cambridge, MA, p 255
223. Yamada K, Takahashi J, Kawabata Y, Okada T, Onihara T (1968) SCP from yeast and bacteria grown on hydrocarbons. In: Mateles RI, Tannenbaum SR (eds) Single cell protein I. MIT Press, Cambridge, MA, p 192
224. Wen CY, Fan LT (1975) Models for flow systems and chemical reactors. Marcel Dekker, Inc, New York
225. Erickson LE, Nakahara T (1975) Process Biochem 10:9
226. Gutierrez JR, Erickson LE (1978) Biotech Bioeng 20:1833
227. Kappeli O, Fiechter A (1976) Biotech Bioeng 18:967
228. Kapelli O, Fiechter A, Finnerty WR (1981) Utilization of n-alkanes by microorganisms: biological aspects of transport. In: Moo-Young M (ed) Advances in biotechnology I. Pergamon Press, Toronto, p 177
229. Blanch HW, Einsele A (1973) Biotech Bioeng 15:861
230. Mimura A, Watanabe S, Takeda I (1971) J Ferment Technol 49:255
231. Kennedy RS, Finnerty WR, Sudarsanan K, Young RA (1975) Arch Microbiol 102:75
232. Aiba S, Haung KL, Moritz V, Someya J (1969) J Ferment Technol 42:211
233. Moo-Young M, Shimizu T, Whitworth DA (1971) Biotech Bioeng 13:741
234. Einsele A, Schneider H, Fiechter A (1975) J Ferment Technol 53:241
235. Yoshida F, Yamane T, Yagi H (1971) Biotech Bioeng 13:215
236. Makula RA, Finnerty WR (1972) J Bacteriol 112:398
237. Prokop A, Ludvik M, Erickson LE (1972) Biotech Bioeng 14:587
238. Kappeli O, Fiechter A (1977) J Bacteriol 131:917

239. Cooper DG, Zajic JE (1980) Adv Appl Microbiol 26:229
240. Ratledge C (1978) In: Watkinson RJ (ed) Developments in biodegradation of hydrocarbons. Applied Science, London, p 1
241. Zajic LE, Panchal CJ (1976) Crit Rev Microbiol 5:39
242. Goldman S, Sabtai Y, Rubinovitz C, Rosenberg E, Gutnick DL (1982) Appl Environ Microbiol 44:165
243. Zuckerberg A, Diver A, Peeri Z, Gutnick DL, Rosenberg E (1979) Appl Environ Microbiol 37:414
244. Gallo M, Bertrand JC, Roche B, Azoulay E (1973) Biochim Biophys Acta 296:624
245. Grund A, Shapiro J, Fennewald M, Bacha P, Leahy J, Markbreiter K, Neider M, Toepfer M (1975) J Bacteriol 123:546
246. Neider M, Shapiro J (1975) J Bacteriol 122:93
247. Watkinson RJ (ed) (1978) Developments in biodegradation of hydrocarbons. Applied Science, London
248. Peterson JA, Basu D, Coon MJ (1966) J Biol Chem 241:5162
249. McKenna EJ, Coon MJ (1970) J Biol Chem 245:3882
250. Lode ET, Coon MJ (1971) J Biol Chem 246:791
251. Finnerty WR, Makula RA (1975) Crit Rev Microbiol 4:1
252. Benson S, Oppici M, Shapiro J, Fennewald M (1979) J Bacteriol 140:754
253. Fennewald M, Benson S, Oppici M, Shapiro J (1979) J Bacteriol 139:940
254. Whitworth DA, Ratledge C (1975) J Gen Microbiol 88:275
255. Kennedy RS, Finnerty WR (1975) Arch Microbiol 102:85
256. Kawamoto S, Tanaka A, Yamamura M, Teranishi Y, Fukui S (1977) Arch Microbiol 112:1
257. Vilenchich R, Akhtar W (1971) Process Biochem 13:41
258. Prokop A, Sobotka M (1975) Insoluble substrate and oxygen transport in hydrocarbon fermentation. In: Tannenbaum SR, Wang DIC (eds) Single cell protein II. MIT Press, Cambridge, MA, p 127
259. Faust U, Prave P (1979) Process Biochem 14:28
260. Kanazawa M (1975) The production of yeast from n-paraffins. In: Tannenbaum SR, Wang DIC (eds) Single cell protein II. MIT Press, Cambridge, MA, p 438
261. Leathen WW (1971) U.S. Patent 3.620.927
262. Silver RS, Cooper PG, Boyle JP (1975) Semi-commercial studies of a petroprotein process based on n-paraffins. In: Tannenbaum SR, Wang DIC (eds) Single cell protein II. MIT Press, Cambridge, MA, p 454
263. Pirnik MP, Atlas RM, Bartha R (1974) J Bacteriol 119:868
264. Dworkin M, Foster JW (1956) J Bacteriol 72:646
265. Goldberg I (1977) Process Biochem 12:12
266. Chandra TS, Shethna YI (1977) J Bacteriol 13:383
267. Quayle JR, Ferenci T (1978) Microbiol Rev 42:251
268. Krug ELR, Lim HC, Tsao GT (1979) Ann Rep Ferment Processes 3:141
269. Hamer G (1979) Biomass from natural gas. In: Rose AH (ed) Economic microbiology, vol 4. Microbial Biomass. Academic Press, London, p 315
270. Colby J, Dalton H, Whittenbury R (1979) Ann Rev Microbiol 33:481
271. Wolfe RS, Higgins IJ (1979) Int Rev Biochem 21:267
272. Higgins IJ, Best DJ, Hammond RC (1980) Nature 280:561
273. Quayle JR (1980) Biochem Soc Trans 8:1
274. Quayle JR (1980) FEBS Lett 117:16
275. Goldberg I (1981) Single-cell protein (SCP) from methanol by bacteria: microbiological and engineering process aspects. In: Moo-Young M (ed) Advances in biotechnology II. Pergamon Press, Toronto, p 419
276. Crawford RL, Hanson RS (eds) (1984) Microbial growth on C_1 compounds. American Society for Microbiology, Washington
277. Goldberg I (1985) Biology of the methylotrophs. In: Demain AL, Solomon NA (eds) Biology of industrial microorganisms. Benjamin/Cummings Publishing Company, Menlo Park, CA, p XX
278. Hanson RS (1980) Adv Appl Microbiol 26:3
279. Wilkinson JF (1970) Symp Soc Gen Microbiol 21:15

280. Whittenbury R, Colby J, Dalton H, Reed HL (1976) Biology and ecology of methane oxidizers. In: Schlegel HG, Gottschalk G, Pfennig N (eds) Microbial production and utilization of gases. Erich Goltze KG, Göttingen, p 281
281. Whittenbury R, Phillips KC, Wilkinson JF (1970) J Gen Microbiol 61:205
282. DeBont JAM, Mulder EG (1974) J Gen Microbiol 83:113
283. Tani Y, Kato N, Yamada H (1978) Adv Appl Microbiol 24:165
284. Foo EL (1978) Process Biochem 13:23
285. Romanovskaya VA, Malashenko YR (1978) Mikrobiologiya 47:120
286. Malashenko YR (1976) Isolation and characterization of new species (thermophilic and thermotolerant ones) of methane-utilizers. In: Schlegel HG, Gottschalk G, Pfennig N (eds) Microbial production and utilization of gases. Erich Goltze KG, Göttingen, p 293
287. Shen G, Kodama T, Minoda Y (1982) Agr Biol Chem 46:191
288. Lawrence AJ, Quayle JR (1970) J Gen Microbiol 63:371
289. De Boer WE, Hazeu W (1972) Antonie van Leeuwenhoek 38:33
290. Shishkina VN, Trotsenko YA (1979) FEMS Microbiol Lett 5:187
291. Loginova NV, Govorukhina NI, Trotsenko YA (1982) Mikrobiologiya 51:38
292. Murrell JC, Dalton H (1983) J Gen Microbiol 129:1197
293. Patt TE, Cole GC, Hanson RS (1976) Int J Syst Bacteriol 26:226
294. Patt TE, Cole GC, Bland J, Hanson RS (1974) J Bacteriol 120:955
295. Wolf HJ, Hanson RS (1979) J Gen Microbiol 114:187
296. Quayle JR (1972) Adv Microb Physiol 7:119
297. Green PN, Bousfield IJ (1982) J Gen Microbiol 128:623
298. Ogata K, Nishikawa H, Ohsugi M (1969) Agr Biol Chem 33:1519
299. Tani Y, Yamada H (1980) Biotech Bioeng 22:163
300. Lee JD, Komagata K (1980) J Gen Appl Microbiol 26:133
301. Tye R, Willets A (1973) J Gen Microbiol 77:1P
302. Large PJ, Peel D, Quayle JR (1962) Biochem J 82:483
303. Kemp MB, Quayle JR (1967) Biochem J 102:94
304. Strom T, Ferenci T, Quayle JR (1974) Biochem J 144:465
305. Quayle JR, Keech DB (1959) Biochem J 72:623
306. Quayle JR, Keech DB (1959) Biochem J 72:631
307. Cox RB, Quayle JR (1975) Biochem J 150:569
308. Shiveley JM, Saluja A, McFadden BA (1978) J Bacteriol 134:1123
309. Van Verseveld HW, Stouthamer AH (1978) Arch Microbiol 118:21
310. Loginova NV, Trotsenki YR (1979) FEMS Microbiol Lett 5:239
311. Quayle JR, Pfennig N (1975) Arch Microbiol 102:193
312. Douthit HA, Pfennig N (1976) Arch Microbiol 107:233
313. Bellion E, Hersh LB (1972) Arch Biochem Biophys 153:368
314. Salem AR, Hacking AJ, Quayle JR (1973) Biochem J 136:89
315. Johnson PA, Quayle JR (1965) Biochem J 95:859
316. Kemp MB, Quayle JR (1966) Biochem J 99:41
317. Anthony C (1975) Sci Prog (Oxford) 62:167
318. Taylor S (1977) FEMS Microbiol Lett 2:305
319. Hazeu W, de Bruyn JC, Van Dijken JP (1983) Arch Microbiol 135:205
320. Fujii T, Asada Y, Tonomura K (1974) Agr Biol Chem 38:1121
321. Fujii T, Tonomura K (1973) Agr Biol Chem 37:447
322. Kato N, Ohashi H, Hori T, Tani Y, Ogata K (1977) Agr Biol Chem 41:1133
323. Yamada H, Tani Y, Kato N, Sakazawa C (1983) Microbiology 139
324. Kato N, Higuchi T, Sakazawa C, Nishizawa T, Tani Y, Yamada H (1982) Biochim Biophys Acta 715:143
325. Sahm H, Roggenkamp R, Wagner F (1975) J Gen Microbiol 88:218
326. Hazev W, Batenburg-van den Vegte WH, Nieuwdorp PJ (1975) Experientia 31:926
327. Van Dijken JP, Otto R, Harder W (1975) Arch Microbiol 106:221
328. Fukui S, Tanaka A, Kawamoto S, Yusahara S, Tanaka A, Osumi M, Imaizumi F (1975) Eur J Biochem 59:561
329. Fukui S, Tanaka A, Kawamoto S, Yusahara S, Terashimi Y, Osumi M (1975) J Bacteriol 123:317

330. Veenhuis M, van Dijken JP, Harder W (1976) Arch Microbiol 111:123
331. Tanaka A, Yasuhura S, Kawamoto S, Fukui S, Osumi M (1976) J Bacteriol 126:919
332. Van Dijken JP, Harder W, Beardsmore AJ, Quayle JR (1978) FEMS Microbiol Lett 4:97
333. Sahm H (1975) Arch Microbiol 105:179
334. Fujii T, Tonomura K (1975) Agr Biol Chem 39:2325
335. Roggenkamp R, Sahm H, Wagner F (1974) FEBS Lett 41:283
336. Colby J, Zatman LJ (1975) Biochem J 148:505
337. Colby J, Zatman LJ (1975) Biochem J 148:513
338. Ben-Bassat A, Goldberg I (1980) Biochim Biophys Acta 611:1
339. Ben-Bassat A, Goldberg I (1977) Biochim Biophys Acta 497:586
340. Sokolov AP, Trotsenko YA (1977) Mikrobiologiya 46:1119
341. Stirling DI, Dalton H (1978) J Gen Microbiol 107:19
342. Ben-Bassat A, Goldberg I, Mateles RI (1980) J Gen Microbiol 116:213
343. Samuelov N, Goldberg I (1982) Biotech Bioeng 24:731
344. Higgins IJ, Best DJ, Hammond RC, Scott D (1981) Microbiol Rev 45:556
345. Wilkinson TG, Topiwala HH, Hamer G (1974) Biotech Bioeng 16:41
346. Goldberg I, Rock JS, Ben-Bassat A, Mateles RI (1976) Biotech Bioeng 18:1657
347. Rock JS, Goldberg I, Ben-Bassat A, Mateles RI (1976) Agr Biol Chem 40:2129
348. Samuelov N, Goldberg I (1982) Biotech Bioeng 24:2605
349. Schneider JD, Wendlandt KD, Bruehl E, Mirschel G, Jenapharm VEB (1983) Z Allg Mikrobiol 23:33
350. Wasungu KM, Simard RE (1981) Baker's yeast production from ethanol by the "Zulauf" process. In: Moo-Young M (ed) Advances in biotechnology II. Pergamon Press, Toronto, p 439
351. Ristroph DL, Watteeuw CW, Armiger WB, Humphrey AW (1977) J Ferment Technol 55:599
352. Prokop A, Votruba J, Sobotka M, Panos J (1978) Biotech Bioeng 20:1523
353. Du Preez JC, Toerien DF, Lategan PM (1981) Eur J Appl Microbiol Biotechnol 13:45
354. Loureiro-Dias MC, Peinado JM (1982) Biotech Lett 11:721
355. Leao C, van Uden N (1982) Biotech Bioeng 24:2601
356. Leao C, van Uden N (1984) Biotech Bioeng 26:403
357. Mor JR, Fiechter A (1968) Biotech Bioeng 10:159
358. Murray DG (1974) Act Rep Res Dir Assoc Mil Food Packaging Syst 26:95
359. McAbee MK (1974) Chem Eng News 9:11
360. Masuda Y (1974) Econ Eng Rev 6:54
361. Masuda Y, Nakanishi N, Sakakura Y (1976) Hydrocarbon Processing 55:113
362. Anon (1976) Eur Chem News 14:53
363. Anon (1980) Eur Chem News 28:20
364. Gaden EL Jr (1974) Substrates for SCP production. In: Davis P (ed) Single cell protein. Academic Press, London, p 47
365. Callihan CD, Clemmer JE (1979) Biomass from cellulosic materials. In: Rose AH (ed) Economic microbiology, vol 4. Microbial biomass. Academic Press, London, p 271
366. Forage AJ, Righelato RC (1979) Biomass from carbohydrates. In: Rose AH (ed) Economic microbiology, vol 4. Microbial biomass. Academic Press, London, p 289
367. Rhodes A, Fletcher DL (1975[2]) Principles of industrial microbiology. Pergamon Press, Oxford
368. Burrows S (1979) Baker's yeast. In: Rose AH (ed) Economic microbiology, vol 4. Microbial biomass. Academic Press, London, p 32
369. Oura E (1974) Biotech Bioeng 16:1197
370. Dellweg H, Bronn WK, Hartmeier W (1977) Kem Kemi 4:611
371. Wang HY, Cooney CL, Wang DIC (1977) Biotech Bioeng 16:69
372. Burrows S (1970) Baker's yeast. In: Rose AH, Harrison JS (eds) The yeasts 3. Academic Press, New York, p 349
373. Reed G, Peppler HJ (1973) Yeast technology. Avi Publishing Company, Westport, CT
374. Harrison JS (1967) Process Biochem 2:41
375. Sato T (1966) Baker's yeast. Korin-Shoin Publisher, Tokyo
376. Anon (1975) Feedstuffs, March 31, 4
377. Rolz C (1975) Utilization of cane and coffee processing by-products as microbial protein substrates. In: Tannenbaum SR, Wang DIC (eds) Single cell protein II. MIT Press, Cambridge, MA, p 273

378. Espinosa R, Maldonado O, Menchu JF, Rolz C (1977) Biotech Bioeng Symp 7:35
379. Jurgensen MF, Patton JT (1979) Process Biochem 14:2
380. Romantschuk H (1975) The Pekilo process: protein from spent sulfite liquor. In: Tannenbaum SR, Wang DIC (eds) Single cell protein II. MIT Press, Cambridge, MA, p 344
381. Drews SM (1975) Sonderheft Berichte der Landwirtschaft 192:755
382. Vananuvat P, Kinsella JE (1975) J Food Sci 40:336
383. Moebus O, Teuber M (1983) General aspects of production of biomass by yeast fermentation from whey and permeate. In: Ferranti MP, Fiechter A (eds) Production and feeding of single cell protein. Applied Science Publishers, London, p 124
384. Moresi M, Sebastiani E (1979) Eur J Appl Microbiol Biotechnol 8:63
385. Moresi M, Nacca C, Nardi R, Palleschi C (1979) Eur J Appl Microbiol Biotechnol 8:49
386. Vrignaud Y (1971) Rev Inst Pasteur Lyon 4:147
387. Meyrath J, Bayer K (1979) Biomass from whey. In: Rose AH (ed) Economic microbiology, vol 4. Microbial biomass. Academic Press, London, p 208
388. Greenberg NA, Mahoney RR (1981) Process Biochem 16:2
389. Prenosil JE, Stucker E, Hediger T, Bourne JR (1984) Biotechnology 2:441
390. Wasserman AE (1960) J Dairy Sci 43:1231
391. Amundson CH (1967) Amer Dairy Rev 29:22
392. Lembke A, Moebus O, Grasshoff O, Reuter H (1975) Sonderheft Berichte der Landwirtschaft 192:571
393. Vringnaud Y (1976) Die österreichische Milchwirtschaft 31:405
394. Moebus O, Kiesbye P, Teuber M (1977) Kiel Milchwirtsch Forschungsber 29:131
395. Skogman H (1976) Production of symba yeast from potato wastes. In: Birch GG, Parker KJ, Worgan JT (eds) Food from waste. Applied Science, London, p 167.
396. Kirk TK (1971) Ann Rev Phytopathol 9:185
397. Nimz H (1974) Angew Chem 86:336
398. Adler E (1977) Wood Sci Technol 11:169
399. Crawford DL (1981) Biotech Bioeng Symp 11:275
400. Detroy RW, Hasseltine CW (1978) Process Biochem 13:2
401. Ban S, Glanser-Slojan M (1979) Biotech Bioeng 21:1917
402. Humphrey AW, Moreira A, Armiger W, Zabriskie D (1977) Biotech Bioeng Symp 7:45
403. Fan LT, Gharpuray MM, Lee YH (1981) Biotech Bioeng Symp 11:29
404. Fan LT, Lee YH, Gharpuray MM (1983) Adv Biochem Eng 23:157
405. Millett MA, Baker AJ, Satter LD (1976) Biotech Bioeng Symp 6:125
406. Han YW (1978) Adv Appl Microbiol 23:119
407. Horton GL, Rivers DB, Emert GH (1980) Ind Eng Chem Prod Res Dev 19:422
408. Ladisch MR, Ladisch CM, Tsao GT (1978) Science 201:25
409. Cowling EB, Kirk TK (1976) Biotech Bioeng Symp 6:95
410. Fan LT, Lee YH, Beardmore DH (1981) Biotech Bioeng 23:419
411. Gharpuray MM, Lee YH, Fan LT (1983) Biotech Bioeng 25:157
412. Rexen F (1983) Principles for pre-treatment of cellulose substrates. In: Ferranti MP, Fiechter A (eds) Production and feeding of single cell protein. Applied Science Publishers, London, p 2
413. Bungay HR (1983) Prospects in the United States for using lignocellulosic materials. In: Ferranti MP, Fiechter A (eds) Production and feeding of single cell protein. Applied Science Publishers, London, p 15
414. Goldstein IS (1976) Biotech Bioeng Symp 6:293
415. Roberts RS, Sondhi DK, Bery MK, Colcord AR, O'Neil DJ (1980) Biotech Bioeng 10:125
416. Knappert D, Grethlein H, Converse A (1980) Biotech Bioeng 22:1449
417. Gaden EL Jr, Mandels MH, Reese ET, Spano LA (eds) (1976) Enzymatic conversion of cellulosic materials: technology and applications. Biotech Bioeng Symp 6. John Wiley & Sons, New York
418. Brown RD JR, Jurasek L (eds) (1979) Hydrolysis of cellulose: mechanisms of enzymatic and acid catalysis. Adv Chem Series 181. American Chemical Society, Washington, DC
419. Lee YH, Fan LT (1980) Adv Biochem Eng 18:102
420. Lee YH, Fan LT, Fan LS (1980) Adv Biochem Eng 17:132
421. Kosaric N, Ng DCM, Russell I, Stewart GC (1980) Adv Appl Microbiol 26:147
422. Ghose TK (1978) Microbial technology in the provision of energy and chemicals from renewable resources, submitted to the Fermentation Commission IUPAC RO 171975

423. Montenecourt BS, Kelleher TJ, Eveleigh DE, Pettersson LG (1980) Biotech Bioeng Symp 10:15
424. Montenecourt BS (1983) Strain improvement for the production of microbial enzymes for biomass conversion. In: Ferranti MP, Fiechter A (eds) Production and feeding of single cell protein. Applied Science Publishers, London, p 30
425. Shoemaker S, Schweickart V, Ladner M, Gelfand D, Kwok S, Myambo K, Innis M (1983) Biotechnology 1:691
426. Shoemaker S, Watt K, Tsitovsky G, Cox R (1983) Biotechnology 1:687
427. Sprey B, Lambert C (1983) Cellulases: delicate exoproteins. Demonstration of multienzyme complexes within the culture liquid of *Trichoderma reesei*. In: Ferranti MP, Fiechter A (eds) Production and feeding of single cell protein. Applied Science Publishers, London, p 50
428. Paice MG, Desrochers D, Rho D, Jurasek L, Roy C, Rollin CF, de Miguel E, Yaguchi M (1984) Biotechnology 2:535
429. Gum EK Jr, Brown RD Jr (1977) Biochim Biophys Acta 492:225
430. Labudova I, Farkas V (1983) Biochim Biophys Acta 744:135
431. Fan LT, Lee YH (1983) Biotech Bioeng 25:2707
432. Lee YH, Fan LT (1982) Biotech Bioeng 24:2383
433. Lee YH, Fan LT (1983) Biotech Bioeng 25:939
434. Woodward J, Whaley K, Zachry GS, Wohlpart DL (1981) Biotech Bioeng Symp 11:619
435. Ghose TK, Das K (1971) Adv Biochem Eng 1:55
436. Howell JA, Stuck JD (1975) Biotech Bioeng 17:873
437. Bisset F, Sternberg D (1978) Appl Environ Microbiol 35:750
438. Ryu DDY, Mandels M (1980) Enzyme Microbiol Technol 9:91
439. Sternberg D, Vijayakumar P, Reese ET (1976) Can J Microbiol 23:139
440. Allen A, Sterberg D (1980) Biotech Bioeng Symp 10:189
441. Klei HE, Sundstron DW, Goughlin RW, Ziolkowski K (1981) Biotech Bioeng Symp 11:593
442. Jain D, Ghose TK (1984) Biotech Bioeng 26:340
443. Takagi M, Abe S, Suzuki S, Emert GH, Yata N (1977) Proc Bioconversion Symp Indian Inst Technol Delhi 551
444. Kilian SG, Prior BA, Lategan PM (1983) Eur J Appl Microbiol Biotechnol 18:369
445. Kilian SG, Prior BA, Potgieter HJ, du Preez JC (1983) Eur J Appl Microbiol Biotechnol 17:281
446. Kilian SG, Prior BA, Pretorius IS, du Preez JC, Venter JJ, Potgieter HJ (1983) Eur J Appl Microbiol Biotechnol 17:334
447. Avgerinos GC, Fang HY, Biocic I, Wang DIC (1981) A novel, single-step microbial conversion of cellulosic biomass the ethanol. In: Moo-Young M (ed) Advances in biotechnology II. Pergamon Press, Toronto, p 119
448. Peitersen N (1975) Biotech Bioeng 17:1291
449. Callihan CD, Dunlap CE U.S. Environmental Protection Agency, Rep SW-24C. PB 203-630
450. Callihan CD, Dunlap CE U.S. Environmental Protection Agency, PB 223-873
451. Han YW (1982) J Ferment Technol 60:99
452. Ek M, Eriksson KE (1975) Appl Polym Symp 28:197
453. Viesturs UE, Apsite AF, Laukevics JJ, Ose VP, Bekers MJ, Tengerdy RP (1981) Biotech Bioeng Symp 11:359
454. Mishra MM, Yadav KS, Kapoor KK (1981) Zbt Bakt II Abt 136:603
455. Kristensen TP (1978) Eur J Appl Microbiol Biotechnol 5:155
456. Aguirre F, Maldonado O, Rolz C, Mench JF, Espinosa R, de Cabrera S (1976) Chemtech 6:636
457. Imrie FKE, Vlitos AJ (1975) Production of fungal protein from carob (*Ceratonia siliqua* L). In: Tannenbaum SR, Wang DIC (eds) Single cell protein II. MIT Press, Cambridge, MA, p 223
458. Forage AJ, Righelato RC (1978) Prog Ind Microbiol 14:60
459. Pirt SJ (1975) Principles of microbe and cell cultivation. Scientific Publications, Blackwell, London
460. Jones RP, Greenfield PF (1984) Process Biochem 19:48
461. Scott R (1983) Design and evaluation of experiments to provide scale-up information from the ICI single cell protein process. In Biotech 83. Online Publications Ltd, Northwood, England, p 235
462. Monod J (1949) Ann Rev Microbiol 3:371
463. Wang DIC, Cooney CL, Demain AL, Dunhill P, Humphrey AE, Lilly MD (1979) Fermentation and enzyme technology. John Wiley & Sons, New York

464. Monod J (1950) Ann Inst Pasteur, Paris 79:390
465. Novick A, Szilard L (1950) Proc Natl Acad Sci 36:708
466. Herbert D, Elsworth R, Telling RC (1956) J Gen Microbiol 14:601
467. Holme T (1962) Adv Appl Microbiol 4:101
468. Malek I, Fencl Z (eds) (1966) Theoretical and methodological basis of continuous culture by microorganisms. Academy of Science, Prague, Czechoslovak
469. Tempest DW (1970) Methods in Microbiol 2:259
470. Pirt SJ (1972) J Appl Chem Biotech 22:55
471. Dean ACR (1962) Proc Roy Soc 13:580
472. Dean ACR, Pirt SJ, Tempest DW (eds) (1972) Environmental control of cell synthesis and function. Academic Press, New York
473. Jannasch JW, Mateles RI (1974) Adv Microbiol Physiol 11:165
474. Cooney CL (1979) Ann NY Acad Sci 79:295
475. Tempest DW, Neijssel OM (1981) The utility of chemostat cultures in studies of microbial physiology. In: Moo-Young M (ed) Advances in biotechnology I. Pergamon Press, Toronto, p 185
476. Goldberg I, Er-el Z (1981) Process Biochem 10:2–8
477. Mateles RI, Battat E (1974) Appl Microbiol 28:901
478. Aiba S, Humphrey AE, Millis NF (1973^2) Biochemical Engineering. Academic Press, New York
479. Herbert D (1961) Soc Chem Ind Monograph 12:21
480. Harrison DEF (1971) Rep Prog Appl Chem 570
481. Ryu DY, Mateles RI (1968) Biotech Bioeng 10:385
482. Young TB, Brady DF, Bungay III HR (1970) Biotech Bioeng 12:747
483. Young TB, Bungay HR (1973) Biotech Bioeng 15:377
484. Harrison DEF, Topiwala HH (1974) Adv Biochem Eng 3:167
485. Koplov HM, Cooney CL (1979) Adv Biochem Eng 12:1
486. Sikyta B (1983) Methods in industrial microbiology. Ellis Horwood Limited, Chichester, England
487. Rokem JS, Goldberg I (1984) The application of continuous culture in the study of the biological utilization of C_1-compounds. In: Dean ACR, Ellwood DC, Evans CGT (eds) Continuous culture 8: Biotechnology, medicine and the environment. Ellis Horwood Limited, Chichester, England, p 251
488. Smirnova ZS, Gorskaya LA, Dibtsov VP, Osokina NV (1977) Culture development conditions as selection factor for various types of bacteria from mixed methane-oxidizing culture in a continuous process. In: Skryabin GK, Ivanov MV, Kandratjeva EN, Zavarzin GA, Trotsenko YA, Nesterov AI (eds) Microbial growth on C_1-compounds. Abstracts of the international symposium. USSR Academy of Science, Puschino, USSR, p 179
489. Bungay HR, Clesceri LS, Andrianas NA (1981) Auto-selection of a very rapidly growing microorganism. In: Moo-Young M (ed) Advances in biotechnology I. Pergamon Press, Toronto, p 235
490. Dibstov VP, Lalov VV, Grigorian AN (1980) Effects of temperature and pH on the growth of methane-oxidizing cultures. In: Tonge GM (ed) Microbial growth on C_1-compounds. Abstracts of the international symposium. Sheffield, England, p 102
491. Snedecor B, Cooney CL (1974) Appl Microbiol 27:1112
492. Minami K, Yamamura M, Shimizu S, Ogawa K, Sekine N (1978) J Gen Appl Microbiol 25:155
493. Hazeu W, Batenburg-van der Vegte WH, de Bruyn JC (1980) Arch Microbiol 124:211
494. Goldberg I, Cooney CL (1981) Appl Environ Microbiol 41:148
495. Goto S, Yamamoto M, Tsuchyiaya M, Kondo O, Okamoto R, Takamatsu A (1978) J Ferment Technol 56:516
496. Kapultsevich YG, Tretyakova VP (1977) Yeast selection on methanol media. In: Skryabin GK, Ivanov MV, Kondratjeva EN, Zavarzin GA, Trotsenko YA, Nesterov AI (eds) Microbial growth on C_1-compounds. Abstracts of the international symposium. USSR Academy of Science, Puschino, USSR, p 166
497. Papoutsakis E, Hirt W, Lin HCL (1981) Biotech Bioeng 23:235
498. Linton JD, Vokes J (1978) FEMS Microbiol Lett 4:125
499. Smith AJ, Moore DS (1977) Bacteriol Rev 41:419
500. Egli Th, Kappeli O, Fiechter A (1982) Arch Microbiol 131:1

501. Egli Th, Kappeli O, Fiechter A (1982) Arch Microbiol 131:8
502. Eggeling L, Sahm H (1981) Arch Microbiol 130:362
503. Eggeling L, Sahm H (1981) Growth of *Hansenula polymorpha* on mixed substrates and regulation of alcohol oxidase synthesis. In: Moo-Young M (ed) Advances in biotechnology I. Pergamon Press, Toronto, p 267
504. Hazeu W, Donker RA (1983) Biotechnol Lett 5:399
505. Dostalek M, Haggstrom L, Molin N (1972) Optimization of biomass production from methanol. In: Terui G (ed) Fermentation technology today. Society Fermentation Technology, Kyoto, Japan, p 497
506. Haggstrom L (1977) Appl Environ Microbiol 33:567
507. Yamane T, Kishimoto M, Yoshida F (1976) J Ferment Technol 54:229
508. Yano T, Kobayashi T, Shimizu (1978) J Ferment Technol 56:416
509. Kuhn HJ (1975) Einfluß der Temperatur auf das Wachstum von *B. cladotenax*. Thesis No 6435, ETH, Zürich
510. Kuhn HJ, Friedrich U, Fiechter A (1975) Eur J Appl Microbiol Biotechnol 6:341
511. Summers RJ, Boudreaux DP, Srinivasan VR (1979) Appl Environ Microbiol 38:66
512. Chalfan Y, Mateles RI (1972) Appl Microbiol 23:135
513. Kim JH, Ryu DY (1976) J Ferment Technol 54:427
514. Tsuchiya Y, Nishio N, Nagai S (1980) Eur J Appl Microbiol Biotechnol 9:121
515. Goldberg I, Sneh E, Battat E, Klein D (1980) Biotech Lett 2:419
516. Nagai S, Nishizawa Y, Aiba S (1973) J Gen Appl Microbiol 19:221
517. Edwards MF, Wilkinson WL (1972) Chem Eng 265:328
518. Goma G, Moletta R, Maugeri F (1981) Importance and usefulness of the content of maintenance in SCP production. In: Moo-Young M (ed) Advances in biotechnology II. Pergamon Press, Toronto, p 453
519. Solomon BO, Erickson LE (1981) Process Biochem 16:44
520. Nagai S, Aiba S (1972) J Gen Microbiol 73:531
521. Nagai S, Mori T, Aiba S (1973) J Appl Chem Biotechnol 23:549
522. Allais JJ, Baratti J (1983) J Ferment Technol 61:339
523. Jara P, Allais JJ, Baratti J (1983) Eur J Appl Microbiol Biotechnol 17:19
524. Levine DM, Cooney CL (1973) Appl Microbiol 26:982
525. Van Dijken JP, Otto R, Harder W (1976) Arch Microbiol 111:137
526. Rokem (Rock) JS, Goldberg I, Mateles RI (1978) Biotech Bioeng 20:1557
527. Nyeste L, Kirchknopf L, Bathory J, Kassem M, Hollo J (1981) Biomass production from methanol by means of a bacterium. 1. Kinetic data. In: Moo-Young M (ed) Advances in biotechnology II. Pergamon Press, Toronto, p 425
528. Stouthamer AH (1977) Energetic aspects of the growth of microorganisms. In: Haddock BA, Hamilton WA (eds) 27th Symposium of the society of general microbiology. Cambridge University Press, Cambridge, England, p 285
529. Stouthamer AH (1979) Int Rev Biochem 21:1
530. Hellingwerf KJ, Lolkema JS, Otto R, Neijssel OM, Stouthamer AH, Harder W, van Dam K, Westerhoff HV (1982) FEMS Microbiol Lett 15:7
531. Van Dijken JP, Harder W (1975) Biotech Bioeng 17:15
532. Anthony C (1978) J Gen Microbiol 104:91
533. Goldberg I, Mateles RI (1975) Appl Microbiol 122:47
534. Papoutsakis E, Lim HC, Tsao GT (1978) AIChE J 24:406
535. Herbert D (1976) Stoichiometric aspects of microbial growth. In: Dean ACR, Ellwood DC, Evans CGT, Melling J (eds) Continuous culture 6. Applications and new fields. Ellis Horwood Publishers, Chichester, England, p 1
536. Payne WJ (1970) Ann Rev Microbiol 24:17
537. Bell GH (1972) Process Biochem 7:21
538. Linton JD, Stephenson RJ (1978) FEMS Microbiol Lett 3:95
539. Pirt SJ (1965) Proc R Soc Lond (Biol) 163:224
540. Verstraete W, Voets JP (1978) Z Allg Mikrobiol 18:135
541. Pirt SJ (1982) Arch Microbiol 133:300
542. Stouthamer AH, Bettenhausen C (1973) Biochim Biophys Acta 301:53
543. Pipyn P, Verstraete W (1978) Biotech Bioeng 20:1883

544. Miura Y, Okazaki M, Komemushi S, Sakata T, Shiroza T, Abana S (1971) J Ferment Technol 57:124
545. Mateles RI (1971) Biotech Bioeng 13:581
546. Cooney CL, Wang DIC, Mateles RI (1969) Biotech Bioeng 11:769
547. Abbott BJ, Clamen A (1973) Biotech Bioeng 15:117
548. Anderson C, LeGrys GA, Solomons GL (1982) The Chem Eng February 43
549. Charles M (1978) Adv Biochem Eng 8:1
550. Liu MS, Branion RMR, Duncan DW (1973) Biotech Bioeng 15:213
551. Haggstrom MH, Dostalek M (1981) Eur J Appl Microbiol Biotechnol 12:107
552. Morinaga Y, Yamanaka S, Yoshimura M, Makinami K, Hirose Y (1979) Agr Biol Chem 43:2453
553. MacLennan DG, Ousby JC, Vasey RB, Cotton NT (1971) J Gen Microbiol 69:355
554. Harrison DEF (1972) J Appl Chem Biotechnol 22:417
555. Harrison DEF (1973) Crit Rev Microbiol 2:185
556. Harrison DEF (1976) In: Rose H, Tempest DW (eds) Advances in microbial physiology. Academic Press, p 243
557. Puhar E, Prave P, Fiechter A (1981) Second european congress of biotechnology. Eastbourne, England, p 84
558. Puhar E, Lorencez I, Fiechter A (1983) Eur J Appl Microbiol Biotechnol 18:131
559. Swartz JR (1979) Ph D Thesis, MIT
560. Swartz JR, Cooney CL (1981) Appl Environ Microbiol 41:1206
561. Pilat P, Prokop A (1976) J Appl Chem Biotechnol 26:445
562. Chen SL, Gutmanis F (1976) Biotech Bioeng 18:1455
563. Battat E, Goldberg I, Mateles RI (1974) Appl Microbiol 28:906
564. Babij T, Ralph BJ, Pickard PAD (1975) The effect of methanol concentration and carbon dioxide on yield of biomass of *Pseudomonas* TB582. In: The Organizing Committee (eds) Microbial growth on C_1-compounds, proceedings of the international symposium. The Society of Fermentation Technology, Kyoto, Japan.
565. Hansford GS, Humphrey AE (1966) Biotech Bioeng 8:85
566. Ryu DDY (1967) Ph D Thesis, MIT, Cambridge, MA
567. Luscombe B (1974) J Gen Microbiol 83:197
568. Brooks JD, Meers JL (1973) J Gen Microbiol 77:513
569. Senior PJ, Windass J (1980) Biotechnol Lett 2:205
570. Sherwood M (1980) Chem Ind 20:921
571. Ashley MHJ (1981) Continuous sterilization of media. A publication of John Brown Engineers and Constructors Limited, England
572. Harrison DEF (1978) Adv Appl Microbiol 24:129
573. PAG, Revised PAG Guidelines (1980) Food Nut Bull 15:59
574. Hong B, Choi NH, Fan LT (1983) Sequential utilization of mixed sugars by *Clostridium acetobutylicum*. In: Reilly PJ (ed) Proceeding of the thirteenth annual biochemical engineering symposium. State University, Iowa, IA, p 1
575. Lamb SC, Garver JC (1980) Biotech Bioeng 22:2097
576. Lamb SC, Garver JC (1980) Biotech Bioeng 22:2119
577. Hamer G, Heden CG, Carenberg CO (1967) Biotech Bioeng 9:499
578. Linton JD, Drozd JW (1982) Microbial interactions and communities in biotechnology. In: Bull AT, Slater JH (eds) Microbial interactions and communities I. Academic Press, London, p 357
579. Harrison DEF (1976) Chemtech 6:570
580. Linton JD, Buckee JC (1977) J Gen Microbiol 101:219
581. Malashenko YR, Romanovskaya VA, Bogachenko VN, Kryshtab TP (1974) Mikrobiologiya 43:290
582. Harrison DEF, Wilkinson TG, Wren SJ, Harwood JH (1975) Mixed bacterial cultures as a basis for continuous production of single cell protein from C_1 compounds. In: Dean ACR, Ellwood DC, Evans CTG, Mellingo J (eds) Continuous culture 6. Application and new fields. Ellis Harwood Publishers, Chichester, England, p 122
583. Cremieux A, Chevalier J, Combet M, Dumenil G, Parlovar D, Ballerini D (1977) Eur J Appl Microbiol 4:1
584. Parshina SN, Kapultsevich YG, Sterkin VE, Glazunov AV (1982) Mikrobiologiya 51:575

585. Jones GL, Loveless JE, Novak AJ (1975) Technical memorandum TM 116. Water Research Centre, Stevenage Herts, England, p 1
586. Otto R, Hugenholtz J, Konings WN, Veldkamp H (1980) FEMS Microbiol Lett 9:85
587. Akaki M (1965) J Ferment Technol 43:365
588. Malashenko YR, Romanovskaya VA, Bogachenko VN, Khotyan LV, Voloshin NV (1973) Mikrobiologiya 42:403
589. Romanovskaya VA, Malashenko YR, Sokolov IG, Kryshtab TP (1976) The competitive inhibition of the microbial oxidation of methane by ethane. In: Schlegel HG, Gotschalk G, Pfennig N (eds) Microbial production and utilization of gases. E. Goltze KG, Göttingen, p 345
590. Drozd JW, Khosrovi B, Down J, Bailey ML, Barnes LJ, Linton JD (1980) Biomass production from natural gas. In: Sikyta BS, Fencl Z, Polacek V (eds) Proceedings of the 7th international continuous culture symposium. Czechoslovak Academy of Science, Prague, p 505
591. Drozd JW, McCarthy PW (1981) Mathematical model of microbial hydrocarbon oxidation. In: Dalton HD (ed) Proceedings of the 3rd international symposium on microbial growth on C_1-compounds. Sheffield, England, p 360
592. Wilkinson TG, Hamer G (1979) J Chem Biotechnol 29:56
593. Evans GH (1968) Industrial production of single-cell protein from hydrocarbons. In: Mateles RI, Tannenbaum SR (eds) Single cell protein I. MIT Press, Cambridge, MA, p 243
594. Horowitz A, Gutnick D, Rosenberg E (1975) Appl Microbiol 30:10
595. Hopwood DA (1981) Scientific American 245:67
596. Mateles RI (1979) Proc GIAM-V 315
597. Baltzis BC, Fredrickson AG (1983) Biotech Bioeng 25:2419
598. Aris R, Humphrey AE (1977) Biotech Bioeng 19:1375
599. Slater JH, Bull AT (1978) Interactions between microbial populations. In: Bull AT, Meadow PM (eds) Companion to microbiology. Longman, London, p 181
600. Rokem JS, Goldberg I, Mateles RI (1980) J Gen Microbiol 116:225
601. Kuenen JG, Boonstra J, Schroder HGJ, Veldkamp H (1977) Microbial Ecology 3:119
602. Kuenen JS, Gottschal JC (1982) Competition among chemolithotrophs and methylotrophs and their interactions with heterotrophic bacteria. In: Bull AT, Slater JH (eds) Microbial interactions in communities I. Academic Press, London, p 153
603. Gottschal JC, de Vries S, Kuenen JG (1979) Arch Microbiol 121:241
604. Dostalek M, Molin N (1975) Studies of biomass production of methanol oxidizing bacteria. In: Tannenbaum SR, Wang DIC (eds) Single cell protein II. MIT Press, Cambridge, MA, p 385
605. Linton JD, Harrison DEF, Bull AT (1975) J Gen Microbiol 90:237
606. MacLennan DG, Ousby JC, Owen TR, Steer DC (1974) British Patent 1.370.892
607. Harrison DEF (1981) Mixed cultures in industrial processes. In: Moo-Young M (ed) Advances in biotechnology I. Pergamon Press, Toronto, p 15
608. Harrison DEF, Harwood JH, Wren SJ (1974) British Patent 1.450.412
609. Wren SJ, Harwood JH, Harrison DEF (1974) Proc Soc Gen Microbiol 2:14
610. Wren SJ, Harrison DEF (1976) Proc Soc Gen Microbiol 4:29
611. Wilkinson TG, Hamer G (1973) J Appl Bacteriol 36:309
612. Wilkinson TG, Hamer G (1972) J Appl Bacteriol 35:577
613. Lee YK (1981) The use of algal-bacterial mixed cultures in the photosynthetic production of biomass. In: Bushell ME, Slater JH (eds) Mixed culture fermentations. Academic Press, London, p 151
614. Hamer G (1982) Biotech Bioeng 24:511
615. Hamer G (1983) Anal NY Acad Sci 413:322
616. Labuza TP (1975) Cell collection: recovery and drying for SCP manufacture. In: Tannenbaum SR, Wang DIC (eds) Single cell protein II. MIT Press, Cambridge, MA, p 69
617. Wang DIC (1968) Cell recovery. In: Mateles RI, Tannenbaum SR (eds) Single cell protein I. MIT Press, Cambridge, MA, p 217
618. Dunnill P, Lilly MD (1975) Protein extraction and recovery from microbial cells. In: Tannenbaum SR, Wang DIC (eds) Single cell protein II. Cambridge, MA, p 179
619. Viehweg H, Schugerl K (1983) Eur J Appl Microbiol Biotechnol 17:96
620. Hedenskog G, Mogren H (1973) Biotech Bioeng 15:129
621. Wiseman A (1969) Process Biochem 4:62
622. Mitsuda H, Yasumoto K, Nakamura H (1969) Chem Eng Prog Sym Series 65:93

623. Berend J, Simovitch E, Ollian A, Rosenberg A (1979) Feasibility study. Tahal Consulting Engineers Ltd, Israel
624. Spicer A (1972) Tropical Sci 13:239
625. Moo-Young M, Daugulis AJ, Chahal DC, Macdonald DG (1979) Process Biochem 14:38
626. Lewis CW (1976) J Appl Chem Biotechnol 26:568
627. Anon (1977) Process Biochem 12:30
628. Anon (1982) New Scientist, May 495
629. Tonge GM (1980) Instrumentation and control in fermentation: the application of computers and computer controlled mass spectrometry. A paper distributed in the 6th International Fermentation Symposium, July, London Canada
630. Viikari L, Linko M (1977) Process Biochem 12:17
631. Solomons GL, Scammell GW (1976) USA Patent 3.937.654
632. Newmark P (1980) Nature (London) 287:6
633. Margaritis A, Wallace JB (1984) Biotechnology 2:447
634. Margaritis A, Sheppard JD (1981) Biotech Bioeng 23:2117
635. Shah YT, Kelkar BG, Godbole SP, Deckwer WD (1982) AIChE J 28:353
636. Prave P (1977) Angew Chem Int Ed Engl 16:205
637. Dobry DD, Jost J (1977) Ann Rep Ferment Process 1:95
638. Weigand WA (1978) Ann Rep Ferment Process 2:43
639. Roels JA (1982) J Chem Technol Biotechnol 32:59
640. Pungor E Jr, Schaefer E, Weaver JC, Cooney CL (1981) Direct monitoring of a fermentation in a computer-mass spectrometer-fermentor system. In: Moo-Young M (ed) Advances in biotechnology I. Pergamon Press, Toronto, p 393
641. Kobayashi T, Yano T, Shimizu S (1981) Automatic control of dissolved oxygen concentration with microcomputer. In: Moo-Young M (ed) Advances in biotechnology I. Pergamon Press, Toronto, p 413
642. Nestaas E, Wang DIC (1981) A new sensor, the "filtration probe", to monitor and control antibiotic fermentations. In: Moo-Young M (ed) Advances in biotechnology I. Pergamon Press, Toronto, p 433
643. Heinzle E, Bolzern O, Dunn IJ, Bourne JR (1981) A porous membrane-carrier gas measurement system for dissolved gases and volatiles in fermentation systems. In: Moo-Young M (ed) Advances in biotechnology I. Pergamon Press, Toronto, p 439
644. Tannen LP, Nyiri LK (1979) Microbial Technol 2:331
645. Armiger WB, Humphrey AE (1979) Microbial Technol 2:375
646. Cooney CL (1979) Biotech Bioeng Symp 9:1
647. Bernard A, Cordonnier M, Lebeault JM (1983) Process Biochem 18:2
648. Zabriskie DW (1976) Ph D Thesis. University of Pennsylvania, Philadelphia
649. Wang NH (1976) Ph D Thesis, MIT, Cambridge, MA
650. Armiger WB (1977) Ph D Thesis. University of Pennsylvania, Philadelphia
651. Nanba A, Hirota F, Nagai S (1981) J Ferment Technol 59:383
652. Woehrer W, Hampel W, Roehr M (1981) Ethanol- and RQ-based computer control in fed batch culture of baker's yeast. In: Moo-Young M (ed) Advances in biotechnology I. Pergamon Press, Toronto, p 419
653. Collmer A, Wilson DB (1983) Biotechnology 1:594
654. Gilkes NR, Kilburn DG, Longsford ML, Miller RC Jr, Wakarchuk WW, Warren RAJ, Whittle DJ, Wong WKR (1984) J Gen Microbiol 130:1377
655. Ho NWY, Stevis P, Rosenfeld S, Huang JJ, Tsao GT (1983) Biotech Bioeng Symp 13:245
656. Lawlis VB, Dennis MS, Chen EY, Smith DH, Henner DJ (1984) Appl Environ Microbiol 47:15
657. Stewart BJ, Leatherwood JM (1976) J Bacteriol 128:609
658. Hitchner EV, Leatherwood JM (1980) Appl Environ Microbiol 39:382
659. Ghosh VK, Ghose TK, Gopalkrishnan KS (1981) Biotech Bioeng 24:241
660. Fennington G, Nevbaver D, Stutzenberger F (1984) Appl Environ Microbiol 47:201
661. Van Uden N, Cabeca-Silva C, Madeira-Lopes A, Spencer-Martins I (1979) Biotech Bioeng 22:655
662. Chakrabarty AM, Friello DA (1974) U.S. Patent 3.923.603
663. Chakrabarty AM (1976) Ann Rev Genet 10:7
664. Portis AL, Boronin AM, Skryabin GK (1983) Appl Biochem Microbiol 19:262

665. Windass JD, Worsey MJ, Pioli EM, Pioli D, Barth PT, Atherton KT, Dart EC, Byton D, Powell K, Senior PJ (1980) Nature (London) 287:396
666. Haber CL, Allen LN, Zhao S, Hanson RS (1983) Science 211:1147
667. Gottschalk G, Hamilton WA, Harder W, de Leevw A, Neijssel OM, Ratledge C (1984) Microbial physiology and biotechnological innovation in the EEC countries, Portugal and Spain. A report of the Deutsche Gesellschaft für chemisches Apparatewesen e. V., Frankfurt/Main
668. Moore AT, Nayudu N, Holloway BW (1983) J Gen Microbiol 129:785
669. O'Connor ML, Hanson RS (1978) J Gen Microbiol 104:105
670. Lindstrom ME (1983) Biotech Bioeng Symp 13:239
671. Lindstrom-O'Connor ME (1983) Microbiology 1:155
672. James AP, Zahab DM (1983) J Gen Microbiol 129:2489
673. Maleszka R, James AP, Schneider H (1983) J Gen Microbiol 129:2495
674. Miyasaka Y, Rha C, Sinskey AJ (1980) Biotech Bioeng 22:2065
675. Stewart GG, Russell I (1977) Can J Microbiol 23:444
676. Russell I, Stewart GG, Reader HP, Johnston JR, Martin PA (1980) J Inst Brew London 86:120
677. Stewart GG, Panchal CJ, Russell I (1983) J Inst Brew London 89:170
678. Stewart GG (1981) Can J Microbiol 27:973
679. Russell I, Stewart GG (1979) J Inst Brew London 85:95
680. De Figueroa LI, de Richard MF, de van Broock MR (1984) Biotechnol Lett 6:171
681. Spencer JFT, Spencer DM (1983) Ann Rev Microbiol 37:121
682. Minami K, Yamamura M, Shimizu S, Ogawa K, Sekine N (1978) J Ferment Technol 56:1
683. Limtong S, Seki T, Kinoshita S, Taguchi H, Kumnuanta J (1981) Production of cellulase by thermophilic fungus isolated in Thailand. In: Taguchi H (ed) Microbial utilization of renewable resources 2. Osaka University, Osaka, Japan, p 64
684. Chavanich S, Nilubol N, Hayashida S (1980) Isolation of thermophilic fungi as cellulase-producers. In: Taguchi H (ed) Microbial utilization of renewable resources 1. Osaka University, Osaka, Japan, p 80
685. Chavanish S, Yoshioka H, Hayashida S (1981) A comparative study of cellulose and xylanase produced by some thermophilic fungi. In: Taguchi H (ed) Microbial utilization of renewable resources 2. Osaka University, Osaka, Japan, p 72
686. Miller TF, Srinivasan VR (1983) Biotech Bioeng 25:1509
687. Updegraff DM (1971) Biotech Bioeng 13:77
688. Moo-Young M (1977) Process Biochem 12:6
689. Stringer DA (1975) ICI protein: manufacture, characteristics, toxicology, and nutritional properties. ICI Report, July
690. Ericsson M, Ebbinghaus L, Lindblom M (1981) J Chem Technol Biotechnol 31:33
691. Calloway DH, Kumar AM (1969) Appl Microbiol 17:176
692. Camargo-Rubio E, Ornelas-Vale A, Casarrubias-Arcos G, Nagai S (1977) J Ferment Technol 55:56
693. Iyengar MS, Baruah JN, Singh HD (1971) Production of single-cell protein from petroleum hydrocarbons. In: Freitas YM, Fernandes F (eds) Global impacts of applied microbiology. University of Bombay, Bombay, India, p 552
694. Duthie IF (1975) Animal feeding trials with a microfungal protein. In: Tannenbaum SR, Wang DIC (eds) Single cell protein II. MIT Press, Cambridge, MA, p 505
695. Prokop A, Ratcliffe HD, Fatayer MI, Al-Awadhi N, Khamis A, Murad M, Bond C, Hamdan IY (1984) Biotech Bioeng 26:1085
696. Atkinson B, Mavituna F (1983) Biochemical engineering and biotechnology handbook. Nature Press, New York
697. Goldberg I, Jensen AP (1977) J Bacteriol 130:535
698. Urakami T, Komagata K (1979) J Gen Appl Microbiol 25:343
699. Casalicchio G, Paoletti C, Bernicchia A, Govi G (1975) Mic Ital 1:21
700. Shinozaki H, Ishida M (1976) Brain Res 109:435
701. Soeder CJ (1977) 11th FEBS Meeting, Kopenhagen
702. Endozien JC, Udo UV, Young VR, Scrimshaw NS (1970) Nature (London) 228:180
703. Sinskey AJ, Tannenbaum SR (1975) Removal of nucleic acids in SCP. In: Tannenbaum SR, Wang DIC (eds) Single cell protein II. MIT Press, Cambridge, MA, p 158

704. Miller SA (1968) Nutritional factors in single-cell protein. In: Mateles RI, Tannenbaum SR (eds) Single cell protein I. MIT Press, Cambridge, MA, p 79
705. Weaver TL, Patrick MA, Dugan PA (1975) J Bacteriol 124:602
706. Makula RA (1978) J Bacteriol 134:771
707. Bird CW, Lynch JM (1974) Chem Soc Rev 3:309
708. De Souza NJ, Nes WR (1968) 162:363
709. Reitz RC, Hamilton JG (1968) Comp Biochem Biophys 25:401
710. Bird CW, Lynch JM, Pirt FG, Reid WW, Brooks CJW, Middleditch BS (1971) Nature (London) 230:473
711. Bouvier P, Rohmer M, Benveniste P, Ourisson G (1976) Biochem J 159:267
712. Tornabene TG, Kates M, Gelpi E, Oro J (1969) J Lipid Res 10:294
713. Kramer JKG, Kushawa SC, Kates M (1972) Biochim Biophys Acta 270:103
714. Kushwaha SC, Pugh EL, Kramer JKG, Kates M (1972) Biochim Biophys Acta 260:492
715. Goldberg I, Shechter I (1978) J Bacteriol 135:717
716. Lee HY, Erickson LE (1984) Biotech Bioeng 26:758
717. Senior PJ, Wright LF, Alderson B (1982) U.S. Patent 4.324.907
718. Schlegel HG, Lafferty RM, Krauss I (1970) Arch Microbiol 71:283
719. Trevelyan WE, Harrison JS (1956) Biochem J 63:23
720. Bressani R (1968) The use of yeasts in human foods. In: Mateles RI, Tannenbaum SR (eds) Single cell protein I. MIT Press, Cambridge, MA, p 90
721. Shacklady CA (1974) Process Biochem 9:9
722. Kapsiotis GD (1978) Food Nutr 4:2
723. Mateles RI (1968) Applications of continuous culture. In: Mateles RI, Tannenbaum SR (eds) Single cell protein I. MIT Press, Cambridge, MA, p 208
724. Schwartzkoff CL, Barford JP (1981) Macromolecular composition of yeast as a function of life cycle. In: Moo-Young M (ed) Advances in biotechnology II. Pergamon Press, Toronto, p 387
725. Holme T (1957) Acta Chem Scand 11:762
726. Wilkinson JF, Munro HLS (1967) The influence of growth limiting conditions on the synthesis of possible carbon and energy storage in *Bacillus megaterium*. In: Powell EO, Evans CGT, Strange RE, Tempest DW (eds) Microbial physiology and continuous culture. Her Majesty's Stationary Office, London, p 173
727. Senior PJ, Beech GA, Ritchie GAF, Dawes EA (1972) Biochem J 128:1193
728. Dierstein R, Drews G (1974) Arch Microbiol 99:117
729. Cooney CL, Wang DIC, Mateles RI (1976) Appl Environ Microbiol 31:91
730. Harder W, Dijkhuizen L (1983) Ann Rev Microbiol 37:1
731. Alroy Y, Tannenbaum SR (1973) Biotech Bioeng 15:239
732. Hou CT, Laskin AI, Patel RN (1978) Appl Environ Microbiol 37:800
733. Kanamaru K, Iwamuro Y, Mikami Y, Obi Y, Kisaki T (1982) Agr Biol Chem 46:2419
734. Linton JD, Cripps RE (1978) Arch Microbiol 117:41
735. Shaechter M, Maaloe O, Kjeldgaard NO (1958) J Gen Microbiol 19:592
736. Ecker RE, Schaechter M (1963) Biochim Biophys Acta 76:275
737. Tempest DW, Hunter JR (1965) J Gen Microbiol 41:267
738. Rosset R, Julien J, Monier R (1966) J Mol Biol 18:308
739. Sykes J, Young TW (1968) Biochim Biophys Acta 169:103
740. Brown CM, Rose AH (1969) J Bacteriol 97:261
741. Neidhardt FC, Magasanik B (1960) Biochim Biophys Acta 42:99
742. Du Preez JC, Lategan PM, Toerien DF (1984) FEMS Microbiol Lett 23:71
743. Trevelyan WE (1976) J Sci Food Agric 27:225
744. Hedenskog G, Mogren H, Enebo L (1970) Biotech Bioeng 12:947
745. Newell JA, Seeley RD, Robbins EA (1975) U.S. Patent 3.867.255
746. Shetty KJ, Kinsella JA (1979) Biotech Bioeng 21:329
747. Lewis PN, Lawford HG, Kligerman A, Lawford GR (1982) Biotech Bioeng 4:441
748. Castro AC, Sinskey AJ, Tannenbaum SR (1971) Appl Microbiol 22:422
749. Zee JA, Simard RE (1975) Appl Microbiol 29:59
750. Bueno GE, Otero MA, Gonzalez AC (1982) Biotechnol Lett 4:145
751. Ohta S, Maul S, Sinskey AJ, Tannenbaum SR (1971) Appl Microbiol 22:415
752. Lindblom M, Mogren H (1974) Biotech Bioeng 16:1123

753. Waslien CI, Calloway DH, Morgen S, Costa F (1970) J Food Sci 35:294
754. Maul SB, Sinskey AJ, Tannenbaum SR (1970) Nature (London) 228:181
755. Cronan JE Jr, Gelmann E (1975) Bacteriol Rev 39:232
756. Farag RS, Khalil FA, Ali LHM (1983) J Am Oil Chem Soc 60:795
757. Eroshin VK, Krylova NI (1983) Biotech Bioeng 25:1693
758. Zhelifonova VP, Krylova NI, Dedyukhina EG, Eroshin VK (1983) Mikrobiologiya 52:219
759. Jigami Y, Suzuki O, Nakasato S (1979) Lipids 14:937
760. Ikemoto S, Katoh K, Komagata K (1978) J Gen Appl Microbiol 24:41
761. Van der Linden AC, Thijsse GJE (1965) Adv Enzymol 27:469
762. McKenna EJ, Kallio RE (1965) Ann Rev Microbiol 19:183
763. Mishina M, Yanagawa S, Tanaka A, Fukui S (1973) Agr Biol Chem 37:863
764. Raymond RL, Davis JB (1960) Appl Microbiol 8:329
765. Davis JB (1964) Appl Microbiol 12:210
766. Dunlap KR, Perry JJ (1968) J Bacteriol 96:318
767. Fredricks KM (1967) Antonie van Leevwenhoek 33:41
768. Makula RA, Finnerty WR (1968) J Bacteriol 95:2102
769. Makula RA, Finnerty WR (1968) J Bacteriol 95:2108
770. Edmonds P, Cooney JJ (1969) J Bacteriol 98:16
771. McElroy R, Averner M (1978) NASA Technical Memorandum 78491
772. Peterson GR, Stokes BO, Shubert WW, Rodriguez AM (1983) Enzyme, Microbiol Technol 5:337
773. Freeland JC, Gale EF (1947) Biochem J 41:135
774. Zaitseva GN, Belozersky AN (1957) Mikrobiologiya 26:533
775. Sakurada T (1966) Keio J Med 15:45
776. Kanazawa T (1964) Plant Cell Physiol 5:333
777. Seoka N (1961) Cold Spring Harbor Symposium on Quantitative Biology 26:35
778. Malofeeva IV, Belyanova LP (1970) Mikrobiologiya 39:82
779. Takahashi J, Kobayashi K, Kawabata Y, Yamada K (1963) Agr Biol Chem 27:836
780. Ertola RJ, Lilly MD, Webb FC (1965) Biotech Bioeng 7:309
781. De Zeeuw JR (1968) Genetic and environmental control of protein composition. In: Mateles RI, Tannenbaum SR (eds) Single cell protein. MIT Press, Cambridge, MA, p 181
782. Alroy Y, Tannenbaum SR (1977) Biotech Bioeng 19:1145
783. Alroy Y, Tannenbaum SR (1977) Biotech Bioeng 19:1155
784. Anon (1984) Biotechnol Newswatch 4:5
785. Okanishi M, Gregory KF (1970) Can J Microbiol 16:1139
786. Momose H, Gregory KF (1978) Appl Environ Microbiol 35:641
787. Shay LK, Wegner EH, Reiter SE (1983) Dev Ind Microbiol 24:305
788. Shay LK, Wegner EH (1984) U.S. Patent 4.439.525
789. Gautier F, Bonewald R (1980) Mol Gen Genet 178:375
790. Kangas TT, Cooney CL, Gomez RF (1982) Appl Environ Microbiol 43:629
791. MacLaren DD (1975) Chemtech 5:594
792. Anon (1983) United Nations University/Protein Advisory Group Revised Guidelines No's 6, 12, and 15. Food and Nutrition Bulletin 5:59
793. Stringer DA (1983) Process engineering of SCP product safety considerations. In: Biotech 83. Proceeding of the international conference on the commercial application and implications of biotechnology. Online Publications Ltd, Norwood, England, p 189
794. Woodward JC, Short DD (1973) J Nutr 103:569
795. Takata T (1969) Hydrocarbon Process 48:99
796. Wenk C (1983) The animal nutritionists dream of a new SCP. In: Ferranti MP, Fiechter A (eds) Production and feeding of single cell protein. Applied Science Publishers, London, p 168
797. Schoch U, Schlatter Ch (1983) Toxological evaluation of SCP produced from whey. In: Ferranti MP, Fiechter A (eds) Production and feeding of single cell protein. Applied Science Publishers, London, p 173
798. Young VR, Scrimshaw NS (1975) Clinical studies on the nutritional value of single-cell protein. In: Tannenbaum SR, Wang DIC (eds) Single cell protein II. MIT Press, Cambridge, MA, p 564
799. Carter GB (1981) New Sci 216, April 23

800. Bull AT (1981) Prospects for biotechnology. In: Microbial aspects of biotechnology. Proceeding of a joint meeting. The Royal Irish Academy, p 2
801. Dasilva EJ, Shearer W, Taguchi H (1982) Biotechnology in development co-operation: A developing country's view. In: Annual reports of international center of cooperative research and development in microbial engineering, 5. Osaka University, Osaka, Japan, p 347
802. Zilinskas R (1984) Biotechnology 2:686
803. Durand A, Arnoux P, Teilhard de Chardin O, Chereau D, Boquien CY, Lavios de Anda G (1983) Protein enrichment of sugar beet pulp by solid state fermentation. In: Ferranti MP, Fiechter A (eds) Production and feeding of single cell protein. Applied Science Publishers, London, p 120
804. Stowell JD, Bateson JB (1983) Economic aspects of industrial fermentations. In: Nisbet LJ, Winstanley DJ (eds) Bioactive microbial products 2. Development and production. Academic Press, London, p 117
805. Schaffeld G, Sinskey AJ, Rha CK (1982) J Food Sci 47:2072
806. King PP (1982) J Chem Technol Biotechnol 32:2
807. Rolz C (1984) IOBB Newsletter, January 6
808. Senez JC (1983) Biofutur 19:39

Subject Index

MIX
Papier aus verantwortungsvollen Quellen
Paper from responsible sources
FSC® C105338

If you have any concerns about our products,
you can contact us on
ProductSafety@springernature.com

In case Publisher is established outside the EU,
the EU authorized representative is:
Springer Nature Customer Service Center GmbH
Europaplatz 3, 69115 Heidelberg, Germany

Printed by Libri Plureos GmbH
in Hamburg, Germany